Oil and Regional Development

Konrad Schliephake
translated by
Merrill D. Lyew

The Praeger Special Studies program—utilizing the most modern and efficient book production techniques and a selective worldwide distribution network—makes available to the academic, government, and business communities significant, timely research in U.S. and international economic, social, and political development.

Oil and Regional Development

Examples from Algeria and Tunisia

Praeger Publishers   New York   London

PRAEGER SPECIAL STUDIES IN INTERNATIONAL ECONOMICS AND DEVELOPMENT

Library of Congress Cataloging in Publication Data

Schliephake, Konrad.
    Oil and regional development.

    (Praeger special studies in international economics
and development)
    Translation of Erdöl und regionale Entwicklung.
    Bibliography: p. 195
    1. Petroleum industry and trade—Algeria.
2. Algeria—Economic conditions.   3. Petroleum industry
and trade—Tunisia.   4. Tunisia—Economic conditions.
I. Title.
HD9577.A42S3413   1977   338.2'7'2820965   76-24367
ISBN 0-275-23910-1

PRAEGER SPECIAL STUDIES
200 Park Avenue, New York, N.Y., 10017, U.S.A.

Published in the United States of America in 1977
by Praeger Publishers,
A Division of Holt, Rinehart and Winston, CBS, Inc.

789   038   987654321

© 1977 by The Institute for African Studies,
Hamburg, Federal Republic of Germany

Printed in the United States of America

This book has several purposes: to give detailed information on regional development and particularly that development based on hydrocarbons, in both Algeria and Tunisia; on a more general level, to show the mechanisms and regional effects generated by the oil industry in different economic systems; and to explain some of the political attitudes of oil-exporting countries in the developing world during what we called the "energy crisis."

Field and literature studies were made by the author during his appointment to the staff of the Institut für Afrika-Kunde (incorporated into the Free Town of Hamburg's foundation German Overseas Institute, formerly Deutsches Institut für Afrika-Forschung e.V.). The administrative bodies of both institutions and especially the former's head of staff, Dr. H. Voss, promoted the study and a publication of the original results in German in 1975.

In North Africa, help came especially from the different oil enterprises, namely SONATRACH, Direction des rélations extérieures (with its regional offices and its bureau in Munich headed by Director Abdelkader Machou); from SITEP, TRAPSA, and SEREPT in Tunisia; from the Algerian Ministry of Industry (Director Z. Fares) and the Tunisian Ministry of Economy; from numerous regional administrative bodies and, of course, from the oil people with their charming and traditional hospitality.

In Tunis and Algiers, embassies of the Federal Republic of Germany gave useful hints, and personal assistance was extended by Chr. Eikenberg and H. Rechenberg. Scientific support came from Prof. H. Mensching, Hamburg, and Dr. A. Arnold, Hanover; very fruitful discussions on the results were held with Dr. Djilali Sari and Prof. Issam El Zaim, both of Algiers.

The cartographic outfit was in the care of P. U. Thomsen, Hamburg, and was updated by Mrs. C. Weiss and K. Gerhard, Würzburg. The liberal and stimulating atmosphere of the Geographical Institute of Würzburg University made it possible to revise and structure the text according to the most recent developments. Translation was done by Th. Degner and Miss Gwyn Ludder, B.A., revised and prepared for publication by Eric Williams, M.A., assisted by the author.

CONTENTS

Page

PREFACE   v

LIST OF TABLES AND FIGURE   x

PART I: ANALYTICAL AND REGIONAL
INTRODUCTION

Chapter

1   OIL, MINING, AND REGIONAL DEVELOPMENT   3

     Regional Development and Analysis   3
     Mining and Hydrocarbons   5
     Oil Industry   6
       Indirect Effects   8
       Direct Regional Effects   11
     Notes   12

2   THE GEOGRAPHIC AND ECONOMIC SETTING OF
     TUNISIA   14

     Agriculture and Labor   15
     Industry and Trade   18
     Oil Industry   21
     Oil Politics   30
     Notes   31

3   THE GEOGRAPHIC AND ECONOMIC SETTING OF
     ALGERIA   34

     Physical Setting and Agriculture   35
     Economy   38
     Oil Industry   45
       History of the Oil Industry   45
       Current Situation and Economic Importance   50
       Oil and Development Strategy   54
     Notes   57

PART II: REGIONAL EFFECTS OF OIL
EXPLORATION AND PRODUCTION

4   EFFECTS OF EXPLORATION   65

    Types of Oil Activities   65
    Labor Demand   65
    Monetary Flows   66
    Notes   69

5   EFFECTS OF OIL PRODUCTION IN TUNISIA   71

    El Borma   71
      Labor Market in Medenine Governorate   72
      Monetary Flows   76
      Social and Regional Behavior   78
      Effects on Regional Economy and Infrastructure   81
    Douleb   82
      Labor Market in Kasserine Governorate   82
      Monetary Flows   83
    Sidi El Itayem   85
    Ashtart Offshore Field   86
    Notes   87

6   EFFECTS OF OIL PRODUCTION IN ALGERIA   89

    Facilities and Problems in the Wilaya Oases   90
    Hassi R'Mel Gas Field   93
      Facilities   93
      Labor Force   94
      Ancillary Services and Supply   95
      Associated Processes   95
    Hassi Messaoud Oil Field   96
      Facilities   96
      Labor Force   97
      Ancillary Services   98
      Settlement and Supply   100
      Associated Processes   102
    In Amenas Oil Field   103
      Facilities   103
      Labor Force   104
      Ancillary Services and Settlement   105
      Associated Processes   107
    Notes   107

Chapter   Page

7   SYNOPSIS: OIL PRODUCTION AND REGIONAL
     DEVELOPMENT   109

     Transport and Administration   109
     Effects on the Labor Market   110
       Monetary Effects   114
       Behavioral Changes within the Labor Force   114
     Effects on Other Sectors of the Regional Economy   116
       Agriculture   116
       Trade and Industry   119
       Settlement   121
       Capital Market   123
       Physical Setting   125
     Notes   125

                 PART III: REGIONAL EFFECTS OF OIL
                    TRANSPORT AND PROCESSING

8   EFFECTS OF OIL TRANSPORT   131

     The TRAPSA Harbor in Skhira, Tunisia   132
       Location and Facilities   132
       Labor Market in Sfax Governorate   133
       Effects on the Labor Market   134
       Monetary Flows   136
       Economic and Social Behavior of the Labor Force   136
       Effects on the Skhira Delegation   137
       Infrastructures   140
     Bejaia, Algeria   140
     Notes   143

9   REGIONAL EFFECTS OF OIL-PROCESSING INDUSTRIES
     IN TUNISIA   144

     Gabes as an Industrial Growth Pole   144
       Economic Problems in the Gabes Governorate   145
       Industrial Facilities in Gabes   147
       Associated Processes   151
     Bizerte   152
     Notes   153

Chapter   Page

10   REGIONAL EFFECTS OF OIL-PROCESSING
      INDUSTRIES IN ALGERIA   156

          Arzew   156
            Facilities of the Oil Industry   156
            Effects on the Labor Market   159
            Associated Processes   159
          Skikda   160
            Facilities of the Oil Industry   161
            Effects on the Labor Market   163
            Associated Processes   165
          Setif   166
            Facilities and Projects of the Plastics Industry   166
            Effects on the Labor Market   168
            Associated Processes   168
          Notes   169

11   SYNOPSIS: OIL PROCESSING AND REGIONAL
      DEVELOPMENT   172

          Effects on Spatial Structures   172
          Effects on Population Patterns   174
          Effects on Regional Economy and Agriculture   177
          Notes   180

      PART IV: THE NATURAL GAS INDUSTRY

12   EFFECTS ON FOREIGN RELATIONS AND COMMERCE   183

          Gas Reserves and Production   183
          Gas Exports   184
          Notes   187

13   IMPACT ON THE DOMESTIC ECONOMY   188

          Internal Consumption   188
          Production of Electricity   189
          Notes   190

14   SUMMARY   192

BIBLIOGRAPHY   195

ABOUT THE AUTHOR   205

LIST OF TABLES AND FIGURE

Table		Page
2.1	Tunisian Employment According to Sectors, 1972	15
2.2	Character of Tunisian Agricultural Acreage	17
2.3	Production of Tunisian Oil Fields, 1966–75	22
2.4	Oil/Natural Gas Pipelines in Tunisia	23
2.5	Investments in the Tunisian Oil Sector, 1969–74	24
2.6	Balance of Foreign Oil Trade in Tunisia, 1961–74	25
2.7	Foreign Currency Inflow to Tunisia, 1973–76	26
2.8	Tunisian Ordinary State Revenues, 1972–76	27
2.9	Duties Paid by Oil-Producing Companies in Tunisia, 1964–75	28
2.10	Investments of Foreign Enterprises in the Tunisian Oil Industry, 1959–70	28
2.11	Jobs in the Tunisian Oil Industry, 1960–68	29
3.1	Areas Differentiated According to Rainfall and Slope Inclination as a Percentage of the Total Area of Northern Algeria	35
3.2	Evolution of Nonagricultural Employment in Algeria, 1966–73	41
3.3	Development of Algerian Industrial Employment in Branches, 1967–74	42
3.4	Algerian Job Offerings, 1969 and 1973	43
3.5	Prospective Development of the Employment Figures in the Algerian Nonagricultural Sectors, 1969–80	44
3.6	Investments of the Algerian Oil Industry, 1952–62	47

3.7   SONATRACH Participation in the Activities of the
        Algerian Oil Industry, 1966-73   50

3.8   Activities of the Algerian Oil Industry: Extent of
        Exploration, Production, and Profits, 1952-76   52

3.9   Part of Oil Income in Algerian State Revenues,
        1971-76   53

4.1   Oil Exploration in Tunisia: Size and Demand for
        Manpower, 1954-70   67

4.2   Oil Exploration in Algeria: Size and Demand for
        Manpower, 1954-73   68

5.1   People Economically Active in Medenine According
        to Sectors, 1966   72

5.2   Change in the Residential Population in Medenine
        According to Delegations, 1956-66   73

5.3   Migration Balances of the South Tunisian Governor-
        ates, 1956-71   74

5.4   Migration of the Southern Tunisian Population,
        1966   75

5.5   Monthly Salaries in Medenine   77

5.6   Population Structure and Annual Construction Applica-
        tions in Medenine   79

5.7   Nonagricultural Jobs in Kasserine, 1973   83

6.1   Rate of Population Growth in Algeria on Wilaya
        Basis, 1966-71   91

6.2   Estimates of the Available Incomes of Active Popula-
        tion in the Wilaya Oases, 1971-73   92

6.3   Origin and Qualification of the Employed in Hassi
        Messaoud, 1971   98

Table		Page
6.4	Origin and Qualification of the Employed in Hassi Messaoud, 1973	99
6.5	Origin and Qualification of the In Amenas Work Force, Production Division, 1973	105
7.1	Origin of the South Algerian Oil Workers in Hassi Messaoud, 1968 and 1973	111
7.2	Origin of the South Algerian Oil Workers in In Amenas, 1973	112
7.3	Population in the Dairas of the Algerian Sahara, 1968	113
7.4	Development of Date Production in Zelfana, 1969-73	118
7.5	Investment Sums Available from Oil Revenues in the Algerian Sahara	124
8.1	Economically Active Persons in Sfax in 1973, According to Branches	134
8.2	TRAPSA Contractors in Skhira	135
8.3	Origin and Behavior of the TRAPSA Employees in Skhira	137
8.4	Employment Provided in Bejaia, 1966	142
9.1	Changes in the Residential Population in the Gabes Delegations, 1956-66	145
9.2	Active Population and Employment in Gabes, 1972	146
10.1	Projects in the Plastics Industry in Setif	167
11.1	Qualifications of the Algerian Industrial Workers in Hydrocarbons, Chemical Industry, and Other Branches, Including Construction Enterprises, 1970 and April 1974	176

Table   Page

12.1   Algerian Natural Gas Reserves   184

12.2   Algerian Export Contracts for Liquefied Natural Gas
         as of December 1975   185

Figure

Oil Production and Oil/Gas Pipelines in Algeria and Tunisia
                                                              frontispiece

1   Myrdal's Process of Cumulative Causation--A Simple
         Illustration   4

xiii

OIL PRODUCTION AND OIL/GAS PIPELINES IN
ALGERIA AND TUNISIA

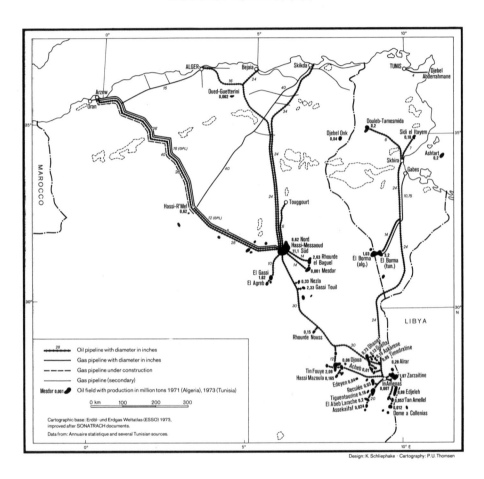

Design: K. Schliephake · Cartography: P.U. Thomsen

ANALYTICAL AND
REGIONAL INTRODUCTION

1

OIL,
MINING, AND
REGIONAL DEVELOPMENT

REGIONAL DEVELOPMENT AND ANALYSIS

Economic development can be interpreted on both a macro-economic (national or international) and a microeconomic (regional) level. These analytical concepts are closely related to economic growth models that may favor isolated, "growth pole with spread effects," or balanced growth development. In developing countries, spread effects are likely to be weaker than the polarization forces because of barriers such as regional subsistence economies and societies, poor transportation, and poor education. Therefore, at least in its early stages, in a liberal economy economic development is likely to widen regional disparities in economic and social standards of living. Further, regionally and nationally unbalanced growth is an inevitable concomitant of economic development.[1]

In this vein, Gunnar Myrdal has drawn a model of a cumulative causation chain (see Figure 1). It shows not only the backward and forward linkages of a newly founded industrial enterprise in a populated but not thoroughly industrialized country but also explains the mechanism of relative concentration.

We cannot dwell on the justification for either isolated, growth pole, or balanced-growth development. However, even though most of the productive factors are optimized in an isolated or growth-pole development, a balanced growth is definitely most beneficial in view of the population that is the agent of change and on which changes act. It is, therefore, generally speaking, the ideal and the ultimate aim of development planners. In order to evaluate the effects of change in the course of social and economic development, microregional analysis according to the Myrdal model is certainly necessary.

Development planners in North Africa agree that national planning that proceeds from the top (central level) to the base (regional level) is inappropriate, inefficient, and difficult to apply. Therefore a regionalization of planning is conditio sine qua non of realistic and integrated development planning, which is in itself based on the results of regional analysis.[2]

For that reason it makes sense to begin with this regional analysis where developmental problems and strategies are concerned, although this is much more difficult than deriving macroeconomic overviews from official statistical data. We will, then, discover the bases of political decisions that, in Third World countries, tend to be generated more and more according to the analyses of regional realities and problems. In addition, although this study has rather practical and informational intentions and is less theory-minded, it shows that macroeconomic and/or microregional geographical analysis can con-

FIGURE 1

Myrdal's Process of Cumulative Causation—A Simple Illustration

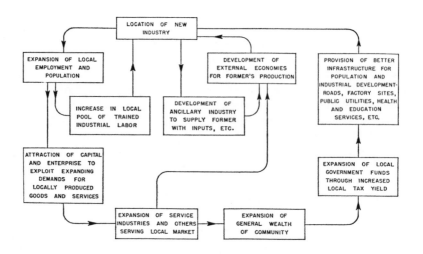

Note: The two subcircles are included to illustrate the ramifications of the cumulative causation process.

Source: D. E. Keeble, "Models of Economic Development," in Models in Geography, ed. R. J. Chorley and P. Hagget (London: Methuen, 1967), p. 258; after G. Myrdal, Economic Theory and Under-Developed Regions (London, 1957).

tribute to the understanding of national and worldwide economic and social development. In the following chapters we try to follow this concept and analyze the micro- and macroregional levels in order to compare their interaction.

MINING AND HYDROCARBONS

Every trade or industrial enterprise, as an economic unit, is closely related to the other activities and characteristics of its location and region and, therefore, has spatial effects. The linkages forward and backward shown by Myrdal also occur within mining activities. The disadvantage of mining, in comparison to other activities that generate development, lies in the fact that the desired development depends on the unpredictable hazards of geology itself. Conditions of geological as well as micro- and macroeconomic prerequisites must be satisfactory in order to give life to the mining industry. Nevertheless, geological surprises and risks are always present.

This leads to the notion of the fugitive and/or speculative character of the mining industry. In fact, in a liberal economy the scattered locations of deposits and the compulsion to an interstate exchange of mining products have made mining heavily dependent on the world market and its mechanism. It becomes compulsory for a management to operate on optimum economic levels in all areas, rarely leaving room for the extraordinary. As a result, one can postulate that, worldwide, the managements of mining enterprises of their respective resource groups are operating on the same principles—of course with certain regard for regional external, natural, and economic factors. Therefore, other things being equal, such activities tend to generate similar spatial effects. These effects can only partly be determined quantitatively in advance as they also depend on a number of noneconomic factors such as physical nature of the deposit, availability of workers, and so forth.

Traditionally, mining in developing countries was most important in that such enterprises often provided the sole source of incoming foreign currency to the state. Since then, several of the nonindustrialized countries with mining industries have gained an increasing amount of control over their raw materials, which brought about a shift in their favor as far as economic decisions are concerned. Not only the financial effects, such as revenues paid to the state from the raw-material-exporting enterprises, but also questions of regional development in the long run, based on the natural resources, become fundamental.

In principle, the oil-mining industry is not very different from other branches of the mining industry, even though it may be less

labor- and more capital-intensive than traditional ore and coal mining. The problems and questions raised in oil-producing countries
are similar to those raised in any country that is dependent on mining
or raw material production. We therefore can apply our considerations from this chapter to oil as well.

Of course, each of the oil-producing countries has its own unique
developmental problems. With the interrelation among population,
crude oil output, and available infrastructures as criteria, the oil-
producing countries of the developing world can be roughly divided
into the following groups:[3]

1. Producing countries with high per capita income, such as
Saudi Arabia, Kuwait, Qater, Abu Dabi, and Libya. Their financial
resources are rapidly increasing, but infrastructural shortcomings
in the material and personnel sectors and the lack of a sufficient nationwide market make it difficult to establish industries; their capital-
absorbing capacity is lower than their revenues.

2. Producing countries with average per capita income, such
as Algeria, Iraq, Iran, and Venezuela. They appear to offer the best
conditions for quick industrial development as they possess an infrastructure capable of sustaining investments, as well as the requisite
capital.

3. Producing countries with low per capita income, such as
Indonesia and Nigeria. Because of their large populations, the high
oil revenues of these countries either give rise to great social disparities and, thus, to political unrest, or the revenues are distributed
relatively evenly and it is difficult to accumulate capital for industrial
investments because of the excessive demands for consumer goods
that are being met chiefly by imports.

Whereas Algeria clearly belongs to the second group, Tunisia
is somewhere between the second and the third. Thus, they both may
serve as models for their categories. With Tunisia and Algeria as
examples, the bases of the political options derived from the reaware-
ness of regional needs and potentialities will be examined here. A
change in priorities thus becomes apparent. This is representative
of many raw-material-producing countries and will certainly influence
their future policies.

OIL INDUSTRY

The oil industry is the most modern branch of the mining business. * From the beginning of its development, it has been an indus-
_____

*In the following chapters "oil industry" is used in the most gen-
eral sense, including natural oil and gas production, transport, and

trial branch closely connected with modern technology. In contrast
to the iron industry, it is too young to carry the burden of historical
problems of location and forms. Until recently it has been directed
almost exclusively by international enterprises, whose managements
operated according to a purely capitalistic point of view without regard
to regional, ethnic, cultural, or other irrational factors. These enterprises were "technologically, pragmatically and internationally
oriented."[4]

Since oil is more than often found in places where it is not needed,
the market for it and its derivatives is worldwide and obvious, and of
course no country can afford to work uneconomically. No exclusively
domestic cycle, protected from competition, would be able to develop
special forms (except perhaps for the Soviet Union during certain periods of recent history). Oil technology and the principles according
to which the oil industry is being conducted are exactly the same in
all lands, for there is only one technologic and economic optimum according to the "international character of the oil."[5] Because of this
single optimum, which is looked for everywhere, it is easy to find
similarities in the oil industry and in its spatial appearances and effects. This would also explain why, after an analysis of the spatial
effects, various producer countries are so concerned with having a
stronger influence over the oil industries of their countries.

The importance of the oil industry is incontestable. Between
1960 and 1973 its contribution to the world energy supply doubled to
57 percent, whereas coal fell to a mere 19 percent.[6] In numerous
regions of the developing world it seems to be the only source of innovations toward the development of a modern industrial society, and
it is a forceful agent in transforming traditional economic and social
structures. The effects of these innovations have often been described
but not evaluated in view of their spatial relevance. The latter approach
is difficult because of the remoteness of oil installations, the deficien-
cies of statistics, especially in developing countries, and the general
lack of quantitative data concerning the most recent economic and social changes in such countries.

Until the much-talked-about "energy crisis," which struck the
world in the fall of 1973, analyses of the national and international oil
and energy policies focused on the national levels and on the mechanism of the international markets. Nationwide studies generally
came to the conclusion that the regional effects of the oil industry, as
well as mining in general, were low and that the monetary flows
toward the state budgets were often not coupled with developmental

---

processing. The more appropriate term might be "hydrocarbons industry" (industrie des hydrocarbures).

actions.[7] The very few existing case studies,[8] on the other hand, pointed out that the mineral oil industry, in its traditional form,

1. lacked integration with the other sectors of the national economy;
2. was, by the logic of its system, prevented from making a useful contribution toward regional economic and social development, and
3. made only a small contribution to the advance of the oil-producing countries under the prevailing systems and economic principles.

Hence, there was a need for integral planning in order to use oil in the interests of the producing countries. Otherwise, a major crisis was bound to occur.

We therefore see that the nationwide financial effects of oil production are not automatically coupled with the regional effects so much desired by the planning authorities. The former effects, which we might also call secondary or indirect, as they only act through an intermediate state budget, are clearly separated from the latter, or direct, regional effects. They will both be analyzed.

Indirect Effects

An economic activity cannot be considered without taking into account the region in which it operates. Workers are always needed to some extent. Additional services are also required, including spatial structures such as traffic facilities and settlements, which in turn will be affected by these new activities. The oil industry belongs to those fields of economic enterprise that, as far as tradition in the developing countries is concerned, has the fewest regional effects. Monetarily, the crude-oil industry is very important, positively affecting the budget and balance of payments, and complementary effects have more regional and sectoral significance.[9]

This view is justified by the behavior of the predominantly foreign oil companies, which almost always monopolize prospecting, production, and the transport of petroleum in the economically less-developed producing countries. Within the structure of the historical dependency between the producer and the consumer country, problems were not visible that would have needed solution through an examination of regional effects. The supply of raw materials to the industrial countries, the consumers, took place without intense participation of the producer countries. Raw material was looked upon as "a present from heaven" by the consumers and they did not study thoroughly how the producer countries might also benefit from the extraction.

Effects of oil, as well as of other export-oriented products, are nevertheless important· If the situation on the world market per-

mits, the producing country can demand an often-important amount of
the producer's rent from the enterprise. In the most advantageous
case, these revenues can be as high as the difference between the local
production costs plus interest on the capital invested and the price of
the product obtainable on the world market.

Because of the demand for energy products and the high cost of
production for alternative sources of energy, oil revenues are high.
From 40 percent in 1973, the amount of oil revenues within the Algerian state budget rose to 46 percent in 1974 and to 59.1 percent
in 1975 (13 billion Algerian dinars [DA] of a total of DA 22 billion).
In Tunisia it was 19.1 percent of the total state budget of 344.6 million Tunisian dinars (TD), in 1975.[10] The case of Libya shows the
limit of the capacity to absorb investments: of the total oil revenues
(in 1974) of $7.6 billion,* only 55.3 percent or $4.2 billion could be
spent in the ordinary investment budget.[11] However, in the eyes of
the producing countries the oil revenues are still regarded as modest,
although they attained, in 1974, the following levels (in billions of
dollars):

Saudi Arabia	20.0
Kuwait	7.0
Libya	7.6
Iraq	6.8
United Arab Emirates	4.1
Algeria	3.7
Iran	17.4
Nigeria	7.0

Source: Petroleum Economist, March 1975.

A look at the relationship between the level of gasoline taxes in
the industrialized countries (especially in Europe) and the oil revenues
of the producer countries shows a certain justification for the producers'
discontent. For instance, before the sharp price rises of the energy
crisis occurred, in the first half of 1973, the Federal Republic of Germany imported crude oil at a total cost of 4 billion deutschemarks
(DM). Approximately half of that sum consisted of revenues received
by the oil-producing countries. By contrast, the Federal Republic
collected DM 14 billion of mineral oil revenues in 1972—a sum considerably higher than the value of the crude oil, and three times more
than the revenues received by the oil-producing countries.

---

*Unless otherwise noted, all dollar figures are U.S. dollars.

Usually the revenues in producing states are at the disposal of the entire country; they are only occasionally attached to certain regions. The Wilaya Oases, where practically all of Algeria's oil is raised, in 1974 received from the Algerian state's total oil revenues of DA 11 billion only DA 15 million or 0.14 percent for investment purposes. With the state budget as a catalyst, there is, regionally, a total break between origin and destination of the money acquired. So, for the geographic and spatial analysis, it makes no difference if the money that is at the state's disposal comes from oil production or other activities. The exclusive "nourishment" of a state budget from a few selected sources is not solely characteristic of oil production. This can also happen through gold smuggling (formerly in Bahrein), aid from foreign governments (Jordan), aid from private foreigners (Israel), or ore mining (Mauretania). In all these cases there is no direct, so to speak, mechanical interrelationship between the industrialization, with respect to the rise in the regional standard of living, and the source of the revenue. Several producing countries, indeed, receive high revenues from their oil production but, until recently, did not have related industries (for example, Libya). Also, there is no "mechanical" relationship between a certain level of state revenue and regional development. The control of the governments over the disposal of these means is not restricted by economic pressures generated by the money sources. The preferences of the state's rulers are unchecked; they might have entirely different preferences. For example, they may like to raise the general standard of living by distributing the money among all citizens, or they might invest the revenues in the industrialized countries on a long-term basis to achieve high-interest gains.

So even the highest state revenues are—from a regional point of view—only secondary contributions toward a genuine economic development. They do not necessarily bring about an improvement in regional and general structures of the economies, nor an organic growth.[12]

The analysis of the monetary effects of the oil industries in producing countries must be juxtaposed with the regional effects and potentials. From this relationship arises the urgent question of capital-absorption capacity.[13] The introduction of surplus foreign currencies from the oil-producing countries into the international capital markets does not save the capital stocks of the producing countries from depreciation, international inflation, and market disturbances. For short terms, productive investment of the capital is possible only in some of the producer countries. In others, the absorptive capacity is overstrained.[14] The creation of an all-Arab Common (capital) Market only solves the problem if, at the same time, sound investment projects are sought in the participating countries.

So far, the revenues from the sale of oil have not only not been incorporated into the integrated development on the regional level, but they have also lost part of their value as a result of the continuous dollar and pound devaluations. The fact that large portions of the profits are invested outside the producer countries proves that the integration of the petroleum sector into a general process of economic development on a local level—which would be the most natural basis for sound investments—has not yet come about. Therefore these profits should be transferred into productive regional investments. Once again, this raises the question of the linkages of the capital derived from oil with the regional primary effects of the oil industry as a scope for the public investor.

Direct Regional Effects

The chapters following will deal almost exclusively with the regional economic effects of prospecting, production, transportation, and processing. They are typical effects, triggered more or less automatically; they can be strengthened or weakened by government planning.

The efforts of the developing countries toward modernization in all fields of the economy, as well as toward industrialization and balanced regional development, show that the primary and secondary effects of the various activities must go hand in hand. Whereas the available capital (that is, secondary effects) raises macroeconomic and political questions such as control and steering of investments and their productivity, problems of the spatial structures, locational preferences, and economic mechanisms affecting the space (that is, primary effects) must be clarified for the sake of those regions that are to be developed.

The impacts that are directly triggered by the oil industry become active on three levels:

1. Activities and technical installations of the oil industry directly change the physiognomy of the landscape. The buildings and installations add visible structures; their description is rather unnecessary, as emphasis should be placed on their spatial dynamics. Technical details are, therefore, only given from the economic viewpoint and for purposes of general information.

2. The second level is comprised of the people who have contact with the oil industry itself. The economic behavior of this social group and its relationship to society can be examined.

3. Considerably more important are those impacts that affect the spatial level and those that contain a constantly increasing self-improvement in their locational quality for industrial activities.

These questions will be dealt with in view of the overall economic and social structures of Tunisia and Algeria in Chapters 2 and 3.

NOTES

1. A. O. Hirschmann, The Strategy of Economic Development (New Haven, Conn., 1958). For an overall view on growth-pole concepts in developing countries, see especially A. R. Kuklinski, ed., Growth Poles and Growth Centers in Regional Planning, UNRISD Regional Planning, vol. 5 (Paris, 1972). The regional realities of balanced and unbalanced growth are shown by Keith Sutton, "Industrialization and Regional Development in a Centrally Planned Economy— The Case of Algeria," Tijdschrift v. Econ. en Sociale Geografie (Amsterdam) 67, no. 2 (1976): 83–94.

2. Moncef Ben Slama, "Régionalisation et planification du développement," Revue tunisienne des Sciences sociales, no. 21 (1970), pp. 91–118.

3. "All Oil Countries Want to Set up Industries," Petroleum Economist (London) 41, no. 9 (1974): 324–26.

4. R. B. McNee, "Functional Geography of the Firm with an Illustration from the Petroleum Industry," Economic Geography 34 (1958): 321–57, with examples from the SOCONY-Concern; see for the "majors" the very critical study by R. R. Odell, Oil and World Power —A Geographical Interpretation (Harmondsworth, 1970).

5. McNee, "Functional Geography," p. 325.

6. Aziz Alkazaz, "Oil Problems Seen from the Arab Point of View," Intereconomics (Hamburg), no. 7 (1975), pp. 206–08.

7. See the Libyan example of "Superaffluence and Underdevelopment" in Rawle Farley, Planning for Development in Libya (New York: Praeger, 1971); and, recently, Konrad Schliephake, Libya— Economic and Social Structures and Development (in German) [Arbeiten aus dem Institut für Afrika-Kunde], vol. 3 (Hamburg, 1976). Nigerian economy is covered by Ludwig Schaetzl, Petroleum in Nigeria (Ibadan, 1969); and Scott R. Pearson, Petroleum and the Nigerian Economy (Stanford, Calif., 1976).

8. P. Marthelot, "La révolution du pétrole dans un pays insuffisamment développé: La Libye," Cahiers d'Outremer (Bordeaux), no. 18, (1965), pp. 5–31.

9. Ludwig Schaetzl, Industrialization in Nigeria: A Spatial Analysis, Afrika-Studien, vol. 81 (Munich: Weltforum, 1973).

10. "Le budget pour 1975," Nouvelles économiques (Algiers), no. 153 (February 1, 1975); and Marches tropicaux, no. 1530 (March 4, 1975), p. 708.

11. For details of the Libyan economy, see Schliephake, Libya.

12. Ties Moeller, <u>Mining and Regional Development in East Africa</u> (in German), Afrika–Studien, vol. 67 (Munich: Weltforum, 1971), p. 13.

13. G. Destanne de Bernis, "Revenus pétroliers et choix d'un éspace de développement," <u>Mondes en développement</u> (Paris), no. 18 (1974), pp. 73–99.

14. Refer to the lecture held during a conference in Kuwait entitled "Surplusses of Arab Oil Producing Countries and Their Investment Possibilities" by the governor of the Algerian National Bank, Abdel Malik Temam, "Excédents financiers—Développement et structures économiques et financiers," <u>El Moudjahid</u> 8 and 9–10 (June 1974).

2

THE GEOGRAPHIC AND
ECONOMIC SETTING OF
TUNISIA

Microregional analysis of economic and industrial activities only becomes meaningful if it is included in a general survey of national economic, social, and physical settings. North Africa's pressing problems have been partially known for a long time. However, other problems only recently became visible and public. It therefore seems to be all the more necessary, especially as the literature in the English language is rather incomplete, to give a comparatively full overall view. This also serves another purpose of the study: to give general regional information about the two countries.

Tunisia is the smallest of the three Maghreb states, with a surface area of 163,000 square kilometers. Its population of 5.62 million (1974), or 35 inhabitants per square kilometer, grew arithmetically to 5.89 million people in 1976. Being a French protectorate until 1956, it has since then become a presidential republic with good relations toward the former colonial power as well as with its Arab neighbors. Since 1974 it has been divided into 18 governorates (formerly 13).[1] The smaller administrative units are called délégations. The economic and political capital of the state is the city of Tunis. Several publications give a good survey of the physical and economic setting of the country.[2]

From the viewpoint of natural resources, Tunisia is certainly the poorest of the Maghreb countries. Nevertheless, the high standard of education that has been reached since its independence in 1956 —22 percent of the population attended school in 1971—is its best asset. This gives human potential to this country with a favorable geographical location in relation to the industrialized countries.

Although the per capita gross national income in 1973 was $460 (which rose, according to Tunisia's Central Bank, to $512 in 1975), which was considerably lower than in neighboring Libya ($3,530) and

Algeria ($570), it had a high effective annual growth rate of 4.4 percent between 1965 and 1973 and more than 8 percent between 1972 and 1976.[3] As in the case of Algeria, it seems that a process of self-sustaining economic growth has been initiated, although this development is heavily influenced by foreign capital and entrepreneurship profiting from low labor costs (see below).

AGRICULTURE AND LABOR[4]

With a total population of 5.4 million in 1972 (approximately 4.4 million in 1962), the potential work force consisted of 1.3 million men (24.4 percent of total) and 1.5 million women (26.7 percent) between the ages of 15 and 64. Taking into consideration students and disabled persons, men requested approximately 1.2 million positions, and women—due to their traditionally much lower degree of activity—approximately 340,000 jobs. This means that at a rate of activity of 21.8 percent (men only) or 28.1 percent (including women), 1.4 million persons (25.9 percent of total population) were employed in one way or another, as Table 2.1 shows.

According to another source, in 1972, 654,000 persons (417,000 men, 237,000 women) were employed in agriculture, where wages range from TD 0.6 to 1.7 per day, of whom 292,000 were wage

TABLE 2.1

Tunisian Employment According to Sectors, 1972

	Permanent	Seasonal	Total
Agriculture			
Men	300,000	250,000	550,000
Women	—	—	250,000
Industry and services			
Men	413,000	84,000	497,000
Women	—	—	92,000
Total			
Men	713,000	334,000	1,047,000
Women	—	—	342,000
Total	—	—	1,389,000

Source: IVe Plan de développement 1973-1976 (Tunis, 1973), p. 119.

earners (101,000 permanent, 141,000 seasonal, and 50,000 occasional workers).[5] In the nonagricultural sectors, 750,000 persons were employed, of whom 230,000 were in industry. The nonagricultural minimum wage was fixed at TD 23.14 per month in January 1974; however, real wages are often below TD 20.

Between 1973 and 1976 the natural growth rate of the population, at present 2.6 percent per annum,[6] will create an annual additional demand for 50,000 jobs, of which 40,000 will be for men. The agricultural sector will not be able to provide the additional jobs; with a supposed increase in availability of agricultural jobs at a rate of 11 percent between 1974 and 1976 (yearly average of 2.7 percent) only the seasonal workers, who up to now have been underemployed and who make up 62 percent of all agricultural workers, can be absorbed.

This means that in comparison to Algeria (see Chapter 3) the differences in quantities disregarded, the agricultural problems tend to be the same. Even though the dualism between modern and traditional sectors cannot be elaborated upon because of a lack of proper information, and although this dualism has been obscured for many years with varying success by the establishment of cooperatives and by state projects for private farmers, one is forced to reflect upon the statistics.

At constant 1966 prices, total agricultural production in 1972 had a value of TD 144.7 million. (At 1974 prices it was worth TD 261.4 million.) Of the first figure, production value of TD 482.3 was attributable to each permanently employed worker (300,000 men), but, if women are included, the figure falls to only TD 180.9. A comparison to the Algerian productivity figures of the private sector (DA 2,230 = TD 234.1), the socialist sector (DA 5,298 = TD 556.3), or of all agricultural workers (DA 3,333 = TD 350), shows how unfavorable this situation is. The ratio between the agricultural work force and the usable agricultural acreage is equally unfavorable. Of the 16.3 million hectares claimed by the Tunisian statistics as state territory, about 9 million hectares are usable, which breaks down into the figures of Table 2.2. Even these figures must be used with care. For instance, cereal-producing areas fluctuated between 1.7 million hectares (1963), 1.0 million hectares (1966), and 1.5 million hectares (1973); in 1975 this figure fell slightly to 1.4 million hectares.

Seven hectares of arable land and 4.3 hectares of orchards were thus available for each permanently employed male. For all male workers (potential work force) only 3.8 and 2.4 hectares, respectively, were available. When female workers are included, the figures dropped to 2.6 and 1.6 hectares, respectively. These numbers are only slightly more favorable than those of the Algerian private sector (4.7 hectares per male worker). Therefore, there is reason to believe that a considerable number of workers are in sur-

TABLE 2.2

Character of Tunisian Agricultural Acreage

	1961/62		1968 (estimate)	
	1,000 Hectares	Percent	1,000 Hectares	Percent
Field produce annually cultivated	3,200	36	2,100	23.2
Orchard and field produce	318	4	{1,310	{14.5
Orchards	992	11		
Wood and halfa grass	1,240	13	1,320	14.6
Meadows and pastures	3,250	36	4,300	47.6
Total	9,000	—	9,030	—

Source: Abdessatar Grissa, Agricultural Policies and Employment: Case Study of Tunisia, Development Centre Studies, Employment series, no. 9 (Paris: OECD, 1973).

plus and will eventually be laid off. If we calculate the actual value of the agricultural production with the help of price indexes of the different agricultural products, the value amounted to TD 185.4 million at the end of 1972, to which each male worker (permanent and seasonal) contributed TD 337.[7] If we compare this with the per capita agricultural productivity of Algeria (TD 350 per worker), 1 percent of the Tunisian workers would be superfluous; when compared with the productivity of the Algerian socialist sector, that would be 16.5 percent of the work force (approximately 90,000 men).

Even if these calculations are of only theoretical value because of the different price structures of agricultural products, and even if absolute figures are missing, they clearly depict an excess of workers in the agricultural sector. Tunisia avoids discussions of this topic, partly because of the lack of solid data. In any case, it can be proved here that the demand for jobs actually far exceeds the official estimates.

In spite of the overstaffing, Tunisian agriculture is not able to supply the population with basic food. Affected by the differing weather conditions of the past years, for instance, cereal production (wheat, barley, oats) had to be supported by imports and fluctuated as follows:[8]

	Domestic Production	Imports
1965	701.0	230.5
1966	429.0	216.9
1967	400.0	418.0
1968	513.0	214.8
1969	417.0	343.9
1970	600.0	321.2
1971	740.0	235.7
1972	1,150.0	208.1
1973	900.0	280.0
1974	955.0	292.5
1975	1,295.0	319.4

The deficit of basic foods is generally balanced by the export of olive oil, fruits, and wine, so that food imports (1962–73) totaling TD 334.5 million (TD 27.9 million per year) were met by exports of TD 407.2 million (TD 33.9 million per year).

The following tasks for Tunisian agriculture are stated in the Four-Year Plan 1973–76:

1. Fulfillment of the domestic demand that is constantly increasing due to the rise in population and the increase of available incomes.

2. Guarantee of the supply for the domestic food processing industry.

3. Fulfillment of the foreign demand, which depends on the price and quality of the domestic production and, in addition, the meeting of the social obligation to stop rural underemployment. [9]

However, these tasks can only be partially carried out.

INDUSTRY AND TRADE

As in Algeria, Tunisian industry is concentrated on the coast, with additional locations of the mining industry (especially phosphates in the Gafsa Governorate) inland.

According to Tunisian employment statistics, which use data from the 1966 census (since then, population and employment numbers have usually been annually extrapolated), it is only partially possible to classify the individual regions according to jobs offered and the degree of activity among the population.

There are two distinct industrial areas that are considerably varied in their structures and significance: first, the important coastal strip with a great variety of job opportunities, and second,

the monostructured border regions in the west.[10] The mining industry is clearly dominant near the borders, while the processing industry is concentrated at coastal locations. Between these two areas is
an industrial vacuum in Central Tunisia.

Along the coastal strip, Tunis is the dominating industrial center, providing 55 percent of all industrial jobs (textiles, 59 percent;
food processing, 64 percent; building material; 71 percent; building
enterprises, 66 percent; and transportation, 72 percent). Bizerte
and Menzel Bourguiba seem to be hindered in their development because of their proximity to this dominant center, whereas the harbor
towns of Sousse and Sfax, located further south, were able to develop
into significant industrial cities with very diversified job offerings.

The investment figures of TD 163.2 million, published by the
Agence de Promotion des Investissements for 1974 under the investment law of April 1972,[11] indicate this trend of unbalanced growth,
which is concentrated in the coastal towns. Tunisian planners, too,
postulate a spatially balanced growth based on the potentials and need
of the different regional entities.[12] However, the polarization forces
in this relatively liberal economy are much stronger than the impact
of the state's steering instruments.

As the distribution of industrial employment is by no means
parallel to the much more dispersed population density, we can differ-
entiate the economic area of Tunisia into the active coastal regions
and the economically passive inland. These regional disparities,
which are with the actual processes of industrialization, also tend to
accentuate the unemployment problem that has existed since colonial
times. This problem actually led to the vast emigration movement
of today.

Actually, under- or nonemployment touches probably 40 percent
of the potentially active population.[13] The figures of the Fourth Development Plan concerning unemployment (130,000 persons) were probably extrapolated from the 1966 census, which stated that 2.1 percent
of the total population is unemployed. This is not a realistic estimate.
Even if we presume, again rather far from reality, that the part of
the population employed in agriculture (57.6 percent) and that employed in other sectors (42.4 percent) remains stable during the plan
period 1973–76, 89,000 jobs must be created for men and 29,000 for
women, while 79,000 persons would have to seek a job abroad.[14]

In fact, from 1971 to 1975 a yearly average of 43,000 new jobs
were created compared with a demand of 55,000. This was far ahead
of the plan, which foresaw a total provision of 119,000 jobs. Actually,
172,000 were created so that temporary emigration to industrialized
countries slowed down.[15] The problem of ever-increasing job demands due to the quick rise in population and consumption standards
nevertheless remains unsolved. Even the Development Plan 1973–76

agrees that a genuine balance between offer of and demand for jobs will not be established in the near future.

The temporary emigration of Tunisian workers to countries north of the Mediterranean and to Libya remains a dire necessity. In 1972 approximately 212,000 Tunisians lived abroad. That figure mounted to 250,000 persons in 1975. The economically active emigrants (50 percent of the total) transferred TD 25.5 million in 1972 and TD 51.7 million in 1974, or TD 20.0 and 34.5 per month, respectively.[16] Next to the income from tourism, this was the most important contribution to the foreign balance of payments. The contributions of the emigrants, which often consisted of consumer and small investment goods, is probably much higher in actuality.[17]

Looking at the official statistics, we see that, in 1973, 42,900 Tunisians found jobs at home and 22,500 abroad, mostly in Libya.[18] A monthly average of 36,564 persons were registered as jobless in contrast to 1,380 positions offered.

Whether tourism, on which many Tunisian hopes are focused, will broaden the base of employment offers seems doubtful. If we look at Algerian and Libyan figures,[19] we get the impression that the economically weaker states, which lack alternative means of economic development, favor an expansion of the tourism industry.

The influx of tourists mounted spectacularly in the 1960s, from 231,087 persons in 1967 to 780,350 in 1971, and declined slightly to 716,003 in 1974. This amounts to a total of 2.03, 6.77, and 5.64 million overnight stays. Gross revenue from tourism rose, however, from TD 71.5 million in 1971 to TD 79.1 million in 1974, at which time it amounted to 7.1 percent of foreign currency receipts. The targets of the Development Plan, which in 1976 forecast 1.5 million tourists entering and a revenue of TD 132 million, will probably not be reached. Thus the goal of job opportunities in tourism to rise from 40,000 to 60,000 positions in 1976 will not be attained.

The elasticity of the tourist demand is influenced by factors independent of the host country, such as the economic situation in the home country and political conditions in the Mediterranean area. The more Tunisia develops its tourist infrastructure, the more sensitive it becomes to fluctuations of the demand. Tourism, therefore, will probably not be the sector in which stable and well-paid employment will be provided. This will rather be the task of a thorough industrialization, coupled with an intensification of agriculture, both of which are based on natural resources and relatively well-qualified manpower. It is important to keep in mind, especially in a comparison with Libya and Algeria, that, due to the relative scarcity of natural assets controlling economic growth, Tunisia's development policies had to be much more prudent than those of the neighboring countries. Although favoring a certain _étatisation_ or state influence on the industrial development, they were more "project-minded" rather than having an integrated approach.[20]

This nonintegrative approach on the national level means that an involvement in the international trade and capital markets, contrary to the Algerian position shown in Chapter 3, is welcomed by the planning authorities. Thus, Tunisia had the highest inflow of foreign capital within the Maghreb countries with $164 million or $34.80 per capita from 1965 to 1967,[21] a trend that continued and was reinforced by the very attractive capital investment law of April 27, 1972.[22]

Keeping in mind that the country and its market potentials are relatively small, an integrated economic development would only be possible within a Maghreb Common Market, as suggested by Robana.[23] But, with the political disparities of the three (or four, if Libya is included) eventual members becoming more and more evident, such a genuine cooperation seems more remote than ever.

OIL INDUSTRY

From the beginning, Tunisian oil production and reserves could never be compared with those of its neighbors.[24] Nevertheless, in the framework of the national economy they are increasingly important, a fact that seems, up to now, not to have been thoroughly realized.

The history of the finding of Tunisia's oil was similar to that of Algeria, except that it was much richer in disappointments. After the first search permits were granted as early as 1894, several more-or-less systematic geologic explorations took place, resulting in a single, unsuccessful test drilling in 1926. Not until 1948 was systematic prospecting resumed, and in 1949 the French company Société de Recherches d'Exploitation du Pétrole en Tunisie (SEREPT)[25] opened the Cap Bon gasfield that has since then supplied Tunis with gas. Between 1950 and 1963, as many as 68 drillings were made (with a total depth of 162,962 meters) by various companies, mainly French, without any economically profitable findings.

Not until October 20, 1964, when the drilling "EB 1" was completed, did the Société Italo-Tunisienne d'Exploitation Pétrolière (SITEP, ENI [Italian] and Tunisian government each 50 percent) find a profitable field in El Borma, which since that time has been systematically developed.

In 1966 and 1967, SEREPT, which until then, although rather unsuccessfully, had undertaken the most extensive explorations in Tunisia, opened the fields of Douleb, Semmama, and Tamesmida. Of the 119 exploratory drillings between 1949 and 1970, only 14 resulted in commercially profitable wells. Having been encouraged by the relatively liberal commissioning of concessions, the search nevertheless continued.

In 1971 the Compagnie franco-tunisienne de Pétroles (CFTP; 50 percent Compagnie francaise des Pétroles, CFP, and 50 percent Tunisian government) opened the Sidi el Itayem field northwest of Sfax. Also in 1971, the French company Aquitaine found, during offshore drillings, the Ashtart deposits 80 kilometers southeast of Sfax. The most recent offshore field is that of Isis, situated near the Kerkennah Islands. After being developed at costs of TD 20 million, production may begin in 1976 with a potential of 1 million tons of oil per year.[26]

The production of hydrocarbons started in 1954, together with the supply of gas to Tunis from the Cap Bon gas field, which, in 1970, was taken over by the state-operated Société tunisienne de l'électricité et du gaz (STEG). Table 2.3 shows its overall development.

As can be seen from the table, oil production of between 5.7 and 6.5 million tons in 1973, once expected by Tunisian officials, was not reached.[27] Excluding the Ashtart oil field where production rose from an expected 1.0 or 1.2 million tons annually to an actual output of 2.3 million tons in 1975,[28] Tunisia's oil reserves are now esti-

TABLE 2.3

Production of Tunisian Oil Fields, 1966-75
(in 1,000 tons)

	El Borma	Douleb and Semmama	Tamesmida	Sidi El Itayem	Bhirat	Ashtart	Total
1966	630	—	—	—	—	—	771
1967	2,227	—	—	—	—	—	2,241
1968	3,081	115	—	—	—	—	3,191
1969	3,527	159	—	—	—	—	3,707
1970	3,936	161	19	—	—	—	4,151
1971	4,006	170	25	—	—	—	4,096
1972	3,600	200		200	—	—	3,975
1973	3,200	200		185	—	200	3,878
1974	2,400	200		140	10	1,200	4,138
1975 (est.)	2,400	250		200		2,500	5,400

Sources: Tunisia: The Development of the Petroleum Industry, E/CN. 14/EP/58, September 24, 1973 (New York: UN Economic and Social Council, 1973); and Central Bank of Tunisia, Statistiques financières, no. 21/22 (1974), p. 45; no. 40 (1976).

TABLE 2.4

Oil/Natural Gas Pipelines in Tunisia

From	To	Diameter in Inches	Length in Kilometers	Capacity in Tons per Year
Cap Bon	Tunis (gas)	4	60	Unknown
El Borma	TRAPSA (oil)	14	115	4,800,000
Douleb	Skhira (oil)	6	165	600,000
El Borma	Connecting pipelines (oil)	2-4	250	ca. 6,000,000
El Borma	Gabes/Ghan-nouche (gas)	10 3/4	298	Initially 182,500,000 m^3
Sidi El Itayem	Skhira (oil)	8	83	

Source: Tunisia: The Development of the Petroleum Industry, E/CN. 14/EP/58, September 24, 1973 (New York: UN Economic and Social Council, 1973), p. 67.

mated to be 80 million tons. With El Borma's production probably falling to 1.7 million tons in 1975-76, it is doubtful that the target of 5.2 million tons, as it was predicted in the Development Plan, will be reached in 1976.[29]

Table 2.4 provides an overview of the existing oil/natural gas pipelines in Tunisia. The oil pipeline from In Amenas to Skhira, which was originally used exclusively for the transport of Algerian oil, will be discussed in Chapter 8. An additional facility is the refinery in Bizerte.

A second refinery that should produce not only for the southern part of the domestic market but also for export is planned for erection in Gabes. Aside from the Bizerte refinery, there is no petrochemical industry in Tunisia now, and as of this writing no other petrochemical projects are known to exist.[30] In an agreement with neighboring Libya, Tunisia has confined itself to the transformation of its domestic phosphates. Thus, a factory for phosphoric acid, of Industries chimiques maghrébines (ICM), began operation in Gabes on February 2, 1972. More about this and other projects in the Gabes region will be found in Chapter 9.

The financial results of this relatively moderate production were not insignificant, being manifested by (1) the capital investments

of foreign companies in the oil sector, and (2) the revenues from oil
export that flowed into the state budget. From the total gross capital
formation (private sector) of TD 59.3 million in 1972 (total investments minus stocks), TD 10.5 million were contributed by agriculture, TD 19.5 million (one-third) by petroleum exploration, and TD
5.7 million by the textile industry. The overall investments in the
oil industry developed as shown in Table 2.5. At the same time, the
private capital transfers to Tunisia amounted (in million TD) to:

	1969	1970	1971	1972
Private credits	18.5	19.7	20.1	15.1
Tunisian participations	10.5	10.3	12.5	20.1

Source: Central Bank of Tunisia, Rapport Annuel
1972 (Tunis, 1974), p. 51.

The foreign interests are mainly found in the oil industry.
Even more distinctly, oil and its derivatives affected the foreign
trade balance as seen in Table 2.6. Oil and oil derivatives raised
their share of the overall export value from 27.7 percent in 1971 and
27.1 percent in 1972, to 30.0 percent in 1973. In 1974, at 35.9 percent, they became the most important export item, followed by foodstuffs (23.7 percent or TD 94.4 million) and minerals (phosphates
and iron ore).
The actual Four-Year Plan 1973-76 foresees a continuous growth
of oil production and industry. During the planning period, both were
expected to contribute 17.2 percent of the foreign currency inflow.

TABLE 2.5

Investments in the Tunisian Oil Sector, 1969-74

Year	Amount in Million TD	Percent of Gross Capital Formation	Gross Capital Formation (public and private sector) in Million TD
1969	8.8	5.9	148.0
1970	8.8	5.8	151.5
1971	13.2	7.6	172.6
1972	20.8	9.7	214.2
1973	25.1	10.7	233.3
1974	34.9	10.5	332.0

Source: Statistiques Financières, no. 36-38 (November 1975),
p. 59.

TABLE 2.6

Balance of Foreign Oil Trade in Tunisia, 1961-74

| Year | Imports | | Exports | | Domestic Use |
	Tons	TD	Tons	TD	Tons
1961	480,109	6,482,408	175	2,311	900
1962	400,819	6,521,331	53	766	6,600
1963	614,980	7,581,526	3	404	85,700
1964	735,653	6,238,263	5,935	60,982	618,000
1965	736,891	6,592,605	10	2,410	681,000
1966	757,592	6,652,486	668,864	4,665,511	820,000
1967	306,605	3,258,511	1,667,492	11,583,042	860,000
1968	27,639	1,341,162	2,334,848	16,358,167	1,050,000
1969	497,394	4,618,846	2,997,819	22,532,186	1,150,000
1970	—	4,700,000	3,170,600	26,000,000	1,200,000
1971	—	4,422,000	2,942,800	31,400,000	—
1972	—	8,772,000	3,699,100	40,700,000	—
1973	—	ca. 10,000,000	3,674,300	53,600,000	—
1974	—	ca. 34,000,000	3,726,800	142,800,000	—

Sources: Tunisia—The Development of the Petroleum Industry, E/CN. 14/EP/58, September 24, 1973 (New York: UN Economic and Social Council, 1973), pp. 77 and 85; Central Bank of Tunisia, Statistiques financières, May–June 1974, p. 53 and November 1975, pp. 54–55; Central Bank of Tunisia, Rapport d'Activité 1972 (Tunis, 1974), pp. 58–61.

TABLE 2.7

Foreign Currency Inflow to Tunisia, 1973–76
(in percent)

Sector	1973–76 Plan	1974 Performance
Products of processing industry*	26.9	14.4
Tourism	26.6	14.8
Mining and energy (phosphate and oil)	17.2	29.3
Transfers of workers from abroad	8.5	9.3
Unprocessed agrarian products*	3.5	15.8
Other transfers	—	1.1

*Figures not comparable.

Source: IVe plan de développement 1973–1976 (Tunis, 1973), p. 56; and Statistiques financières, nos. 36–38 (November 1975), pp. 51 and 60.

Due to the rise in oil prices, they actually contributed 29.3 percent as shown in Table 2.7.

During the planning period,[31] investments of TD 222.5 million, that is 18.8 percent of the TD 1,194 million total, are being assigned to the mining and energy investment sectors. Of this, TD 110.3 million are to be invested in oil and oil-processing industries, namely, TD 64 million for prospecting, 23.5 million for construction of refineries, and 22.8 million for the development of existing fields. The private sector is liable to contribute 40.6 percent.

The annual growth of the energy sector is supposed to reach 6.7 percent, as a result of the increase in oil production from 4.0 million tons in 1972 to 5.0 million tons in 1976. The annual output production of the refineries will remain constant at 1.1 million tons. So, the share of oil among the exports stands at 29.1 percent in the plan; during the previous period it reached 26.3 percent.

The rather conservative estimates of the Development Plan 1973–76, based on the precrisis prices (1972: TD 10.5 per metric ton; 1973: TD 14.2 per ton), were exceeded by far because of the price rises following the energy crises of 1973. Thus, average prices per ton of crude oil reached TD 36.7 in 1974 and then fell to an average of TD 29.7 for the first eight months in 1975.[32] With a rather stable oil production, 1974 saw an export value of oil that was 2.7 times that of the preceding year. Estimates of the plan (oil revenues of TD 51.5 million) were thus widely surpassed.

With the rise in oil revenues, the country theoretically would have been able to become independent of additional capital inflows that, until 1973, were the only means to counterbalance the notoriously negative balance of payments (1972: minus TD 23 million; 1973: minus TD 65 million). During the plan period 1973-76, foreign contributions were expected to be TD 120 million annually, but net capital inflows only rose from TD 49 million in 1972 to TD 86 million in 1974. With foreign contributions still coming in, oil surpluses have been a valuable asset to Tunisian economy and have allowed domestic planners to tackle certain developmental problems with ease for the first time since independence.

In 1975, oil revenues contributed 19.1 percent of the total state budget. Table 2.8 shows the development of the state's revenues in general. Compared with the relevance of oil in foreign trade and public finance, no real importance at the regional level can be stated. Certainly, the taxes paid by the oil companies, consisting of taxes proper and a 50 percent royalty, derived from the posted price, show an increasing amount, according to Table 2.9.

No information is available concerning the oil companies' expenses incurred within Tunisia. The companies were authorized to spend as much as 70 percent of the investment sums (see Table 2.10) outside of the country (salaries for foreign experts, duty-free imported equipment). The average annual investments of TD 10.9 mil-

TABLE 2.8

Tunisian Ordinary State Revenues, 1972-76
(in million TD)

Source of Revenues	1972	1973	1974	1975	1976
Direct taxes	35.5	46.2	53.6	82.7	84.9
Indirect taxes	115.8	122.3	119.5	149.0	178.6
Other revenues	41.7	39.6	40.1	43.6	59.9
Benefits from royalties and state participations	5.6	11.4	33.4	69.3	69.1
Total	202.6	219.5	247.2	344.6	385.5

Source: "Le budget tunisien," Maghreb-Machrek, no. 62 (1974), pp. 18-20; G. Pierre, "Les budgets de l'Algérie, du Maroc et de la Tunisie," Maghreb-Machrek, no. 72 (1976, pp. 39-54.

TABLE 2.9

Duties Paid by Oil-Producing Companies in Tunisia, 1964–75
(in million TD)

Year	Amount Paid	Year	Amount Paid
1964	3.2	1969	23.9
1965	3.0	1970	26.5
1966	6.7	1971	30.0
1967	9.4	1972–74	—
1968	21.1	1975	65.8

Sources: Tunisia: The Development of the Petroleum Industry,
E/CN. 14/EP/58, September 24, 1973 (New York: UN Economic
and Social Council, 1973), p. 95; Central Bank of Tunisia, Rapport
annuel 1972; budget figures, 1975.

lion during the decade 1962–72 are expected to reach a total of TD
110.3 million during the Four-Year Plan, TD 73.3 million of which
are to be derived from private enterprises. The distribution (in million TD) is as follows:

64.1   for petroleum exploration (TD 47.4 mil-
                        lion contributed by foreign oil companies
                        according to the contract terms);

TABLE 2.10

Investments of Foreign Enterprises in the Tunisian Oil Industry,
1959–70
(in million TD)

Year	Amount	Year	Amount
1959	1.89	1965	15.5
1960	0.04	1966	15.8
1961	0.18	1967	12.8
1962	Unknown	1968	13.6
1963	Unknown	1969	9.7
1964	5.2	1970	8.9

Source: IVe plan de développement 1973–1976 (Tunis, 1973),
p. 252.

22.7   for the further development of existing
                    fields;
23.5   for the refining sector; of which TD 22.2
                    million are contributed for the construction
                    of new refineries.

In comparison to the large sums of money that were spent in
Tunisia, the number of jobs offered by the oil companies remained
small (see Table 2.11). Very little benefit on the regional domestic
level was derived from these expenditures. Considering the current
Tunisian economic policy, a strong increase in job offers cannot be
counted on for the future. Of the 120,000 jobs that shall be created
during the Fourth Plan, only 800 (that is, 0.7 percent) are to be contributed by energy production branches.[33]

Certainly far more people than those considered in the statistics
are more or less directly employed by the oil industry. The actual
employees of the oil industry are supplemented by all those "contracting enterprises" that carry out auxiliary services (painting, constructing, gardening, and so on), and also by many who are employed only
for a short time, mainly for prospecting and drilling purposes. It is
impossible to give the exact size of this group, but nevertheless we
will approximate it in Chapter 4.

TABLE 2.11

Jobs in the Tunisian Oil Industry, 1960-68

Year	Number of Existing Industrial Units	Number of Permanently Employed	Number of Seasonally Employed	Total Number Employed
1960	4	436	—	436
1961	5	623	—	623
1962	5	631	—	631
1963	6	753	—	753
1964	6	1,037	10	1,047
1965	6	983	42	1,025
1966	6	1,059	25	1,084
1967	5	988	38	1,026
1968	4	1,023	73	1,096

Source: Tunisia: The Development of the Petroleum Industry,
E/CN. 14/EP/58, September 24, 1973 (New York: UN Economic and
Social Council, 1973), p. 98.

OIL POLITICS

Up to the time of writing, Tunisia is neither a member of the Organization of Petroleum Exporting Countries (OPEC) nor of the Organization of Arab Petroleum Exporting Countries (OAPEC), although there were plans to join in 1975. Even though the country generally accepts the price recommendations of the producers' cartel, its oil and general economic policies can be called rather liberal.[34] Concession agreements between the state and foreign companies for exploration purposes can be transferred into production concessions. The state of Tunisia holds a 50 percent share if commercially profitable deposits are found within the concession area. In addition to duties of 15 to 17 percent from the turnover and 50 to 55 percent of the net profits, the companies are obliged to offer the government up to 60 percent of the production to cover domestic demands.[35] So far, the state intends no further interference into the affairs of the companies where the state has a rather "quiet" share.

The stable political and economic atmosphere has resulted in an open and liberal cooperation between Tunisia and the Italian, French, North American, and German prospecting and extracting companies. This has served to encourage their activities. Recent exploration indicates additional profitable oil deposits in the coastal regions, which, with the short distance to Europe, results in easy sales at favorable prices.

Without doubt, the Entreprise Tunisienne d'Activités Pétrolières (ETAP), founded in accordance with the law of March 10, 1972, will have to be given more consideration in the future. Tunisian Prime Minister Hedi Nouira ascertained, in September 1972, that foreign investors would still have an important place within the petroleum economy. But the Tunisian government will pay close attention to the sound use of its natural resources in investment and regional planning. Unlike its neighboring countries, Liyba and Algeria, the few economic and natural resources of Tunisia do not allow a "politics of power" in its relationship with private foreign enterprises. The especially high exploration risks in Tunisia have kept the state from intervening strongly. Thus, Tunisia remains wide open to foreign investments.[36]

The future role of the national ETAP has not yet clearly emerged.[37] Generally sharing the same location and personnel with the Energy Department of the Ministry of Economics, it has taken no concrete and spectacular actions up to the time of writing. According to the law of March 10, 1972, it should conduct research concerning the oil industry; train Tunisian specialists in the various branches of the oil industry; participate in industrial, commercial, financial, and other activities that are related to the oil and natural gas industry.[38]

First, the institution's purpose is to develop a new oil legislation. Moreover, it will, as a rule, be the Tunisian partner of the foreign oil companies. We certainly cannot state that the Tunisian regional development policies are based on oil and natural gas as energy supply and raw material.[39] The reason for this might be because the commercially developed oil deposits are relatively recent, or that both neighbors are building up petrochemical industries that must be seen as competitors. It might also be possible that the planners for the economy simply were not aware of the fact that hydrocarbons offer a strong basis for meaningful industrialization.

Therefore, oil and natural gas reserves have had only moderate effects on the Tunisian South. In 1962, before the discovery of its energy deposits, Gabes had already been planned as an industrial growth pole, but the construction progressed slowly until 1972. In contrast to the Algerian oil-shipping harbors, Skhira, for example, was not able to attract a processing industry. The decision concerning the location of the country's second refinery was uncertain for a long time and Gabes was not agreed upon until 1974. It becomes apparent, when compared with the very elaborate Algerian oil politics, that the Tunisian oil industry is not yet completely integrated into the national development planning.

Only recently has the country become aware of the potentials of its petroleum and natural gas reserves and their profitable use in regional development. This is a result of the so-called energy crisis that pointed up the importance of oil throughout the world. On the other hand, it is also the result of observations of developmental strategies and plans of the neighboring countries of Libya and, above all, Algeria.

So far, almost exclusively labor-intensive enterprises, such as the textile industry, have been introduced into the country. But now Tunisia also prefers those industries whose technological demands and achievements can become the driving force of development. Therefore the state raised the question of why the oil companies exploiting the Tunisian deposits did not fulfill Tunisia's desire to process the crude oil where it is found. Receiving TD 241 million, or 41 percent of the total state investments in 1976, industrial investment will only now become fundamental in contrast to former years.[40]

NOTES

1. See the definitive Law 74-47 dated June 6, 1974. A map of the new administrative units is shown in R. Bolz and K. Eitner, Institutions of Development Planning in Libya, Documentation Service Africa, Series B, no. 3 (Hamburg: Institute for African Studies, 1974). As all regional data actually available are still based on the old limitations, the latter are generally used.

2. See, in addition to the literature cited in the following chapters, Ghazi Duwaji, Economic Development in Tunisia: The Impact and Course of Government Planning (New York: Praeger, 1967); and H. Mensching, Tunesien (Darmstadt: Wissenschaftliche Länderkunden, vol. 1, 1968); as well as, recently, Reinhardt, Bolz, Tunisia—Economic and Social Structures and Development (in German), Arbeiten aus dem Institut für Afrika-Forschung, vol. 6 (Hamburg, 1977).

3. World Bank Atlas, 1975.

4. Sources, if not otherwise stated, are IVe Plan de développement 1973-76 (Tunis, 1973), especially pp. 119-24; more recent figures from Central Bank of Tunisia, Statistiques Financières, no. 40 (May 1976).

5. M. Seklani, "Le champ d'intervention du salaire minimum s'élargit à l'agriculture," Conjoncture (Tunis), no. 4 (1974), pp. 24-26.

6. As forwarded by the Development Plan. Meanwhile a growth rate of 2.4 percent is indicated. See Industrial Promotion in Tunisia (Tunis: Agence de Promotion des Investissements, 1975).

7. Central Bank of Tunisia, Statistiques Financières, nos. 21 and 22 (May and June 1974), pp. 40-41.

8. Ibid., no. 40 (May 1976).

9. IVe Plan, pp. 192-93.

10. Tunisia—Industrial Work Force, According to Branches, 1966, designed by I. Schilling-Kaletsch.

11. API, Rapport annuel 1974 (Tunis, 1975).

12. Moncef Ben Slama, "Régionalisation et planification du développement," Revue tunisienne des Sciences sociales, no. 21 (1970).

13. Elie Cohen-Hadria, "Perspectives économiques tunisiennes," Maghreb-Machrek, no. 71 (January-March 1976), pp. 9-10.

14. IVe Plan, pp. 119-28.

15. Cohen-Hadria, "Perspectives économiques tunisiennes."

16. IVe Plan, p. 98; and Central Bank of Tunisia, Statistiques Financières, November 1975, p. 60.

17. Konrad Schliephake, "Libyen als Bezugspunkt mediterraner Arbeiterwanderungen—Beispiele aus Tunesien und Malta," Orient (Opladen) 15 (1974): 112-15.

18. Conjoncture, no. 3 (1974), p. 13.

19. Algeria 1973: 250,210 foreigners entered; Libya: 165,700 foreigners entered, of which 79 percent were from Arab states. As a regional example of impact of tourism, see G. J. Tempelman, "Tourism in South Tunisia: Developments and Problems in the Djerba-Zarsis Region," Tijdschrift voor Economische en Sociale Geografie (Amsterdam) 66 (1975): 21-34.

20. As stated already by Samir Amin, L'économie du Maghreb (Grands documents, 25, Paris, 1966), pp. 43-44.

21. United Nations, Industrial Development Survey, vol. 4, ID 83 (New York, 1972), p. 111.

22. There are numerous publications of the investment-promoting organization, Agence de Promotion des Investissements, Tunis, since 1972.

23. Abderrahmane Robana, The Prospects for an Economic Community in North Africa (New York: Praeger, 1973).

24. See the semiofficial presentation, Tunisia: The Development of the Petroleum Industry, E/CN. 14/EP/58, September 24, 1973 (New York: UN Economic and Social Council, 1973).

25. French companies and a 23.8 percent share by the Tunisian state.

26. Marchés tropicaux, no. 1530 (March 7, 1975), p. 708.

27. "Petrole," Europe France Outre-Mer, no. 513 (49/1972), pp. 40-42; and in Information économique africaine (Tunis), no. 7 (1973), pp. 32-33.

28. "Le Gisement d'Ashtart," ELF Bulletin mensuel d'information (Paris), no. 12 (1973), pp. 3-11, with technical details.

29. For the following, see also the rather unprofitable study, H. Gilbert, "Le petrole en Tunisie," Maghreb-Machrek, no. 58 (July/August 1973), pp. 37-41.

30. Tunisia, pp. 66-69.

31. IVe Plan, pp. 73-75, 83, and 93.

32. Central Bank of Tunisia, Statistiques Financières, November 1975, p. 54.

33. IVe Plan, p. 125.

34. Legal bases are the decrees from December 13, 1948; from January 1, 1953; the Law 58-36 from March 15, 1958; and the laws from March 10, 1972.

35. Tunisia, pp. 95-96.

36. See the interview with Prime Minister Hedi Nouira in "Tunis—deuxième décennie de développement," Europe France Outre-Mer (Paris), no. 513 (1972), pp. 6-9.

37. For ETAP, see "Tunisie—une nouvelle puissance pétrolière?" Jeune Afrique (Paris), no. 732 (January 17, 1975), p. 29.

38. Gilbert, "Le Pétrole en Tunisie."

39. Henri Madelin, Pétrole et politique en Méditerranée occidentale, Cahiers de la fondation nationale des sciences politiques, 188 (Paris, 1973), pp. 192-96.

40. Cohen-Hadria, "Perspectives économiques tunisiennes."

3

THE GEOGRAPHIC AND ECONOMIC SETTING OF ALGERIA

Algeria, today a Democratic Popular Republic, has been integrated politically and economically into Western Europe for a long time. Having been a French colony and afterward part of the French homeland since 1832, it became impregnated superficially by the French language and culture. However, it always remained part of the Arab-Islamic world, and it gained independence in 1962 after a bloody liberation war waged since 1954. Today it is one of the decolonized states that are, geographically and economically, very close to Europe. It is a natural and favorite interlocutor and a connecting link with the Afro-Asian countries of the Third World.

Algeria is one of the biggest countries of Africa (second only to the Sudan) with a total surface area of 2.38 million kilometers. However, only 3 percent of it can be agriculturally used, the rest being mountains and the Algerian part of the Sahara Desert.

The population, in mid-1973, was estimated at 14.7 million; the actual growth rate of 3.5 percent annually will lead to a total of 16.3 million inhabitants in mid-1976.[1] Due to a rather high population and a still-marked disparity between a large traditional sector with low productivity and the more productive modern sector of the economy, gross natural product per capita was $570 in 1973. In comparison with other oil-producing countries such as Libya ($3,530), this is rather low, but it is higher than that of neighboring countries like Morocco ($320) and Tunisia ($460). An annual growth rate of 4.3 percent (1965-73) is rather favorable and gives the impression that the first steps toward a self-sustaining growth in the productive sectors have been made.

A host of scientific studies, most of them in the French language, display the multitude of problems and potentials of the Algerian state. Unfortunately, no handbook or manual in the English language is available.[2] Therefore, the following survey on agriculture and industry is rather detailed and goes beyond the level of industry proper.

PHYSICAL SETTING AND AGRICULTURE

The agricultural assets and potentials of the land are closely related to the physical setting, especially the precipitation and relief, as shown in Table 3.1.

Consequently, only 25 percent of the total area of Northern Algeria of 28.5 million hectares, or 3 percent of the total of the state territory, can be cultivated. This includes the coastal regions and those parts of the Tell Mountain Range where the inclination does not technically hinder cultivation. In absolute figures, 6.8 million hectares are considered agriculturally useful land, of which about 4 million are actually used, the rest being fallow.[3]

A more detailed analysis shows that the quality of the land decreases from north to south. In addition, prior to independence the best land belonged to French settlers and is now managed by worker committees. Their production is still rather export oriented and labor extensive. The disparities between the modern "socialist" and the traditional, private farming sectors are among the most urgent problems of Algeria.

The entire rural population was estimated, in 1973, at 9 million people (57 percent of total population), of which 6.5 million depended on agricultural activities. Of all of these persons, only 540,000 were

TABLE 3.1

Areas Differentiated According to Rainfall and Slope Inclination as
a Percentage of the Total Area of Northern Algeria
(excluding the Sahara)

| Slope Inclination in Degrees | Precipitation by Millimeters and Type | | | | |
	600 Humid	400–600 Subhumid	300–400 Semiarid	300 Arid and Desert	Total
0–12.5	6.2	16.2	23.9	28.5	74.8
12.5–25	6.9	5.3	1.9	1.5	17.5
Over 25	4.6	0.5	0.8	0.5	6.4
Total	17.7	22.0	26.6	30.5	

Source: Calculated from Statistique agricole, no. 7 (December 1968), pp. 10–11. (Mistakes caused by statistical gaps.)

permanently, regularly, and gainfully employed, that is, 15.2 percent
of the rural and 21.1 percent of the agricultural population. These
figures grew only by 9 percent from 13.9 percent and 19.4 percent,
respectively, in 1966.[4] Taking the average national rate of employment, the private agricultural sector (including agrarian revolution
installations) accounted for a potential work force of 1.73 million
men in 1973. It disposed of 6.24 million hectares of farmland (cereals
and so on) and 0.55 million hectares of land with special cultures (vine-
yards, orchards, olive groves).[5] Thus, 3.6 hectares of arable land
and 0.3 hectares of special cultures fall to each potential worker. In
comparison to the economically more advanced countries, it can be
clearly seen that there is a very high surplus of workers in agriculture, especially in the traditional sectors. It is, therefore, unrealis-
tic to suppose that the creation of employment in agriculture (and the
part of the population to be employed in the agriculture) will progress
in proportion to the total population. However, Algerian planning
suggests a decrease in the percentage of agricultural workers among
the total population only from 58.3 percent to 54.6 percent between
1969 and 1980.[6]

The actual decrease must be sensibly higher. In 1969, 7 hectares of agriculturally utilizable area fell on each statistically recorded
worker (not each potential worker) in the socialistic sector—in the
private sector it was 4.7 hectares.[7] If each worker in the private sector had 7 hectares at his disposal—which would mean that he could
perhaps earn an income comparable to that of a worker in industry or
in the socialist sector—348,500 persons at the 1969 level (with the agricultural acreage remaining constant) would no longer be needed. At
the 1973 level (theoretical work force potential in agriculture) with
the ratio between the socialist and the private sector being the same,
this amount would rise to 540,000 persons. We still have those occasional workers who are included among the socialist sector figures
who work only part of the year and undoubtedly belong to the covert
unemployed. Similar results are found in the comparison between
the current and potential productivity of these two agricultural sectors.

In the socialist sector mainly colonially introduced cultures (industrial cultures, citrus fruits, wine, fresh vegetables) and cereals
(31.5 percent of the total value of cereal production) are grown,
whereas the private sector supplies the country with traditional food
(cereals, vegetables, olives, dates), and, especially, animal products
(approximately 96 percent). At present, the value of the agricultural
production, as calculated for the 1970/71 period (one year), amounts
to DA 4 billion,[8] of which 38 percent originated in the socialist sector. Foreign trade of foodstuffs was, until 1970, positive, due to
wine exports amounting to DA 450 to 500 million. In 1971, 1972, and
1973, however, it showed a deficit of DA 501,775, and 190 million,

respectively. In 1970/71, 674,600 tons of cereals at a local value of DA 332.7 million had to be imported.[9]

It will be one of the main tasks of the Algerian government to reduce the food deficit that has existed since 1958 and was caused by the increase in population and nutritional standard. For example, the index of food production per capita of the population fell from an average of 100 points in 1961–65, to 79 points in 1971, to 87 points in 1972, and to 77 points in 1973.[10]

It makes sense that an increase in agricultural production is not achieved by raising the number of workers, as this would mean a subsequent rise in the relatively high food prices, or the workers, at least those in the former private sector, would have to be underpaid, perhaps even more than they are now. The present value, for instance, of the cereal production amounts to approximately DA 123.5 per hectare annually in the private sector and to DA 171.7 per hectare annually in the socialist sector. If we only assimilate the productivity of the former to the latter, this would result in an additional cereal production of DA 224.1 million, so that the cereal deficit (at local prices) would be reduced to DA 108.6 million.

Although hypothetical in many respects, the question could be further elaborated if one looks at the total annual productivity per worker of the two sectors: socialist sector, DA 5,298; private sector, DA 2,230.

If—again very hypothetically—the private sector would reach the same productivity as the socialist sector, the former would require no more than 441,200 workers to reach current output, and an additional 50,000 workers might be needed in order to make up for the present foreign trade deficit in the food and foodstuffs branch (including wine exports). This hypothetical work force contrasts with 1976 figures, 1,048,235 persons agriculturally active, so that, according to our calculations, a work force of approximately 500,000 persons would be superfluous. This number even exceeds the potential number of agricultural workers who could be laid off on account of the available arable land. In fact, in the course of the agrarian revolution, which intends a gradual assimilation of agricultural wages and salaries paid in the industrial sector, a substantial number of agrarian workers will become jobless who have so far been living covertly unemployed at the minimum subsistence level.[11] The Algerian agrarian revolution can succeed only if the productivity of each single agricultural worker increases. This, however, automatically brings about more layoffs in the work force. In conclusion, the industrial sector is forced to create more jobs to compensate for this and the rising population as well. Particularly in developing countries, a sound industrialization must depend on natural resources. Aside from the agricultural potential, the mineral resources are thus of decisive importance. In Algeria's case, hydrocarbons are by far the most important resource.

On the other hand, industrialization makes sense only if it is finally based on modernization and mechanization of national agriculture. The real weakness of the Arab oil-producing countries is their incapacity to provide enough food for a mushrooming population. The reasons for their growing deficits are manifold. Most of the colonial regimes saw no importance in modernizing traditional agriculture, but superposed a modern agricultural sector operated by expatriates (for example, the French colonies in the Maghreb), which was capital-intensive, labor-extensive, and export-oriented.[12] Wine-growing in Algeria, the "poisoned heritage of colonialism" to the Muslim population, is such an example.

Since the "energy crisis," the few surplus food producers became increasingly aware of their possible role vis-a-vis the OPEC countries and are behaving more and more as monopolists.[13] Whereas in Tunisia the prices of exported oil rose by 374.5 percent from 1971 to the middle of 1975, unit prices for meat imports mounted by 140 percent, cereals by 203 percent, and sugar by 533 percent.[14] The Algerian state budget for 1975 saw the growth of food subventions from DA 1.9 billion in the previous year to DA 3.2 billion, or 21.1 percent of the whole expenditure. According to Algerian data, from 1972 to 1974 wheat prices multiplied by five from $60 to $320, rice by eight and sugar by four. The monetary efforts in subsidizing could not be maintained, so food prices rose and state subventions fell to DA 2.2 billion in the 1976 budget.[15]

It seems that the Arab countries have to work out, in addition to their "oil strategy," a parallel "nutrition strategy"; this will be much more difficult to realize, but just as important as the former if a real economic and political independence is aimed at.

Although this task has been recently acknowledged, there are no such spectacular agrarian investment and modernization programs as can be found in Libya, for example.[16] In Algeria, during the Five-Year Plan 1973-77, no special emphasis is placed on agricultural development, which receives only DA 16.72 billion or 15.2 percent of total investments. In any case, Algerian planners are becoming gradually aware of the mounting problem. The pros and cons of certain agricultural projects, especially in the Sahara, are demonstrated and discussed in Chapter 7.

ECONOMY

During the colonial period the country's economy evolved along a dual scheme. Those areas favorable to agriculture and the coastal regions were included in the colonial production and marketing processes. However, they did not offer additional room for likewise quickly expanding Islamic population (which increased from 2.3 million people in 1865 to 9.6 million in 1960). The mountain and steppe regions remained stagnant, if they did not have extraordinary natural resources—this being worsened by the pressure of the population growth. This contrast became more distinct with the development of modern industries that, in the beginning, processed agricultural products for exportation to the mother country, and later also produced simple consumer goods for the domestic market (so, quite naturally, they were also situated on the coast). This process of differentiation caused the breakdown of the precolonial economy that was weakly differentiated because of the generally lacking interregional exchange possibilities; and it was mainly collectively oriented, particularly in agriculture, often described in postcolonial literature as idyllic and harmonic. The subsequent self-sustaining, cumulative process described by Arnold, [17] which aggravated regional disparities, resulted in an industrial concentration on a few developmental poles along the Mediterranean coast, whereas during the precolonial period—at least in Algeria and Morocco—many centers were located inland. Today, 40.7 percent of all industrial jobs (389,723 in 1974, or 448,197 including banking and transportation) including building trade and oil industry (or 70.6 percent excluding building trade and oil industry) are still concentrated in Algiers. The Wilayas Oran, Tlemcen, Annaba, and the Oases, because of the oil industry, also have an above-average industrialization rate of more than 21 jobs per 1,000 inhabitants. In seven out of 13 Wilayas (Saida, Mostaganem, El Asnam, Tiaret, Medea, Setif, and Batna) the job offerings in industry per 1,000 inhabitants are below 50 percent of the weighted national average. Disregarding the oil and construction industries, even ten Wilayas do not reach 50 percent of the national average of industrial job offerings. If our statistical data were more detailed than that available on the Wilayate basis, [18] regional differentiation and disparities could be depicted even more poignantly.

The Algerian government is trying to overcome these regionally unbalanced developments. Above all, this should be achieved through "Special Programs" whose task is to improve the infrastructure and to accelerate industrialization. Within this scheme almost DA 8 billion were invested in ten of the least-developed Wilayates, as listed here: [19]

Oases (decision of November 24, 1966); DA 600 million credit line; improvement in health sector, land reclamation, and road construction. Actually, only DA 300 million were spent, an amount of 400 million is given in Chapter 7. The discrepancy stems from different sources.

Aures (Batna) (decision of February 24, 1968); DA 400 million credit line; water supply, soil meliorization, industry and trade.

The Great Kabylia (decision in Tizi Ouzou of October 23, 1968);
    credit line of DA 550 million; agriculture, forestry, industry,
    and tourism. Actually, DA 830 million were spent.
Titteri (decision of June 2, 1969); credit line of DA 1 billion; 992 mil-
    lion actually spent on water supply, small-scale industry, infra-
    structure (school, health, roads).
Setif (decision of October 24, 1970); credit line DA 1 billion; an actual
    amount of 717 million invested in industrialization and moderniza-
    tion of agriculture.
Tlemcen (decision of June 25, 1970); a credit line of DA 1 billion for
    investments in electrification, natural resources, industry, fish-
    ing, and trade.
Saida (decision of 1971); DA 1.104 billion for industry, education,
    creation of jobs.
El Asnam (decision of October 9, 1972); DA 1.111 billion for infra-
    structure, agriculture, education.
Constantine (decision of 1973); credit line DA 1.5 billion, an actual
    amount of 1.082 billion spent.

Additional programs for the regions (Dairates) of Telagh, Aflou, Tis-
semsilt, and Oued Riou total another DA 783 million. Referring to
these programs, President Boumedienne stated: "The success of our
revolution and the guarantee for its continuity depend on the development of the handicapped regions and the ending of the economic disparities between rich and poor parts of our country."[20]
    The basis of all of these projects is the fact that the present
and future population can probably be fed by a steadily intensified agriculture, which, however, will by no means afford employment for
the population. The number of agricultural workers who will have to
be laid off was calculated here (even though very hypothetically).
This number will be reinforced by the growth of that group that is
from the beginning not connected with the agricultural sector.
    Between 1966 and 1971 the population grew from 12,102,000 to
14,643,700 at an annual growth rate of 3.5 percent (figures from 1972
and 1973 show a growth rate of 3.25 and 3.28 percent).[21] It was cal-
culted that by mid-1974 the population would have reached approximately 16 million and will probably pass the 20 million mark between
1980 and 1982.[22]
    Until recently, the natural population growth had been slowed
down by the workers' emigration, especially to France, which brought
the effective growth down to 2.55 percent per year (excluding emigration losses). However, emigration to France came to a halt for political reasons in 1973 and, due to the slowdown in the general economic
development in other industrialized countries of Europe, it has now
almost totally ceased. Algeria has one of the world's highest prolifera-

tion rates, with its natural growth rate thus approaching the effective rate.[23]

The opportunities of agriculture to employ and nourish the inhabitants were shown before. Employment outside of agriculture developed from 1966 to 1973 as shown in Table 3.2. A more detailed survey (see Table 3.3) shows which branches of nonagricultural activities are promoters and participants of the country's economic development.

With a presently estimated population growth of 550,000 persons annually and the assumption that only 20 percent of the population will be economically active, as many as 110,000 jobs will have to be created annually. Most of these jobs will have to be in the secondary sector, as agriculture would probably even lay off manpower. However, outside the agricultural sector, no more than a yearly average of 68,500 jobs was created between 1966 and 1973.

The theoretical manpower supply of 1.1 million persons (20 percent of the nonagricultural workers) to the industrial and other nonagricultural fields was met in actuality by the number of job offerings (not always corresponding to other statistics) shown in Table 3.4.

Performance figures (Table 3.3) show that the projections in Table 3.4 have actually been implemented, as a nonagricultural work force potential of 1.22 million people compares with 1.21 million jobs in 1973. But this only means that additional demands due to population growth are satisfied. For the traditional structural under- and unemployment in the Maghreb countries, no remedy has yet been found.

TABLE 3.2

Evolution of Nonagricultural Employment in Algeria, 1966–73

Sector	Persons Employed (in 1,000s)		Percent Growth 1966–73
	1966	1973	
Industry	100	242	142
Building	70	190	171
Transport	50	76	52
Services	140	180	28
Commerce	190	225	18
Administration	180	297	65
Total	730	1,210	65

Source: Dj. Sari, "L'évolution de l'emploi en Algérie," Maghreb-Machrek, no. 69 (1975), p. 42.

TABLE 3.3

Development of Algerian Industrial Employment in Branches, 1967–74
(in absolute figures)

Branch	1967	1970	1971	1972	1974	Percent Growth 1971–74 (Oct. 31)
Mining	12,129	14,760	17,252	15,585	15,414	-11.9
Oil and natural gas	7,276	16,240	16,584	23,310	43,934	170.5
Food and foodstuffs	20,177	28,032	28,126	32,231	33,756	20.0
Textile industry	7,659	26,245	29,035	27,477	32,342	11.4
Leather and shoes	7,342	5,828	6,822	7,355	7,009	2.7
Chemical industry	6,305	7,411	8,284	8,578	13,294	60.5
Building materials	7,474	8,404	12,611	12,014	16,187	28.4
Iron and steel production	3,952	9,024	9,380	13,163	15,656	66.9
Metal working	15,325	22,575	22,450	24,804	42,332	88.6
Lumber, cork, furniture	7,764	8,876	8,375	7,733	9,891	18.1
Paper, printing	5,931	5,535	5,762	5,733	7,551	31.0
Remaining producing industries	1,205	2,133	2,031	4,799	5,589	175.2
Construction business	62,450	106,925	112,100	131,723	151,958	35.6
Electricity, gas, and water supply	5,759	6,344	6,476	6,476	9,943	53.5
Banking, insurance, realty business	7,134	9,031	9,132	12,000	17,752	94.4
Transport and communications	29,722	32,743	33,580	38,570	43,195	28.6
Total	207,604	310,106	328,000	371,551	465,803	42.0

Sources: Dj. Sari, "Problemes démographiques algériens," Maghreb–Machrek, no. 63 (1974), pp. 32–42; and "Les résultats (partiels) de l'equête emploi et salaires de 1974" (Algiers: Sécrétariat d'Etat au Plan, 1975).

TABLE 3.4

Algerian Job Offerings, 1969 and 1973

Sector	Number of Jobs
Producing trade	135,000
Construction business	70,000
Transport, trade, services	360,000
Public employees (including teachers, and so on)	275,000
Total 1969	840,000
Producing trade (including construction)	170,000
Transport	18,000
Trade and services	30,000
Public employees (mainly teachers)	47,000
Total 1973	1,105,000

Sources: Personal calculations from La Situation de l'Emploi et des Salaires 1970 (Algiers: Sécrétariat de l'Etat au Plan, 1970), pp. 44–47; L'Algérie en chiffres 1962–1972 (Algiers, 1972), p. 13.

The overall estimate, that 30 to 50 percent of all men are without work (1972: 2 million, of which 600,000 registered),[24] is underlined by the calculations of Tiano, who states that the Algerian development plan 1970–73 estimates an under–unemployment rate of 23.7 percent for urban and 51.3 percent for rural areas.[25] This would mean that 150,000 new jobs would have to be created annually.

In any case, the creation of nonagricultural jobs has to be more intensive than it has been. This has been recognized by Algerian development planners as the country's principal problem. By 1973, 300,000 jobs were added to the 1 million nonagricultural jobs of 1969. An additional 450,000 are to be created by 1977 according to the Four-Year Plan.[26]

The long-term development is as shown in Table 3.5. These jobs have to be created above all in the production industry, which recognizes the increasing domestic demands as well as those of the world market. Logically, they must be based on natural resources. Aside from the agricultural potential, hydrocarbons, that is, oil and natural gas, are by far the most important in Algeria.

As shown in the following sections and in Chapter 11, an industrialization of this kind has been initiated in Algeria. It is not the au-

TABLE 3.5

Prospective Development of the Employment Figures in the Algerian Nonagricultural Sector, 1969–80

Branch	1969	Percent of 1969 Total	1973	Percent of 1973 Total	1980	Percent of 1980 Total
Energy–water supply	5,595	0.6	6,300	0.6	11,600	0.7
Mining, quarrying	13,327	1.5	16,000	1.4	22,000	1.3
Iron industry	5,288	0.6	10,700	0.9	14,200	0.8
Iron processing and electrical engineering	15,982	1.9	32,000	2.8	71,000	4.1
Building materials	10,125	1.2	20,000	1.8	36,000	2.0
Textile industry	26,350	3.0	33,300	2.0	40,500	2.3
Leather industry	6,800	0.8	7,800	0.7	9,200	0.5
Chemical industry	7,350	0.9	9,500	0.8	24,500	1.4
Various other industries	17,000	2.0	23,000	2.0	32,000	1.8
Food processing industry	23,400	2.7	30,000	2.6	49,000	2.8
Petroleum industry	15,500	1.8	20,500	1.8	30,000	1.7
Construction business	85,000	9.9	170,000	15.0	290,000	16.7
Trade	200,000	23.2	230,000	20.3	330,000	19.0
Transport	60,000	7.0	78,000	6.9	132,000	7.6
Services	100,000	11.6	120,000	10.6	190,000	10.9
Education	69,000	8.0	86,400	7.6	141,000	8.1
Health services	24,000	2.8	28,000	2.5	41,000	2.3
Administration	178,000	20.6	210,000	18.6	256,000	14.17
Total	862,627	—	1,131,500	—	1,740,000	—

Source: B. Dellouci and A. Bouisri, "La population active et l'emploi," Dossiers documentaires, no. 19–20 (Algiers, 1972), p. 64. Inaccuracy because of rounding.

thor's intention to discuss whether the concept of "industrializing industry," which places emphasis on heavy industry in order to stimulate economic growth in an integrated national economy, is successful.
For instance, most of the public enterprises show a deficit, except
for SONATRACH (Société Nationale pour la Recherche, la Production,
la Transformation et la Commercialisation des Hydrocarbures) (surplus of DA 500 million) and Air Algérie. In 1974 the other public enterprises needed subsidies of DA 3.5 billion from the state budget,
of which the steel industry alone received DA 2 billion.[27] Nevertheless, the growth and productivity of industrial installations give the
positive impression of an economic "take-off" being attained or not
being far away.

OIL INDUSTRY

In the following chapters the Algerian oil industry will be examined with regard to its historical development, its present importance to the Algerian economy, and its part in the elaboration of a
developmental strategy in Algeria. The literature used for this purpose is not listed completely. For further listings consult the bibliography Oil in Africa, completed in consultation with the author (see
footnote on first page of the bibliography to this volume).

History of the Oil Industry

The history of the oil industry began in 1915 with several,
though unprofitable, drillings on the slopes of the Atlas Mountains.
In 1949 the first relevant oil field was found in Oued Guetterini, approximately 100 kilometers southeast of Algiers. At first its annual production was more than 100,000 tons. However, this well has been dry
since 1954.

According to several French sources, it had already been suspected in 1930 that the most important oil fields of Edjeleh and Hassi
Messaoud existed[28] and that Anglo-American oil companies prevented
prospecting in those areas. Nevertheless, the intense activities of
French geologists between the wars did not succeed in finding additional deposits. Thus, there were those who posited that the Sahara
was "dry" (in both senses of the word), even in the above-mentioned
fields. They believed that only small layers of sediment covered the
thick pre-Cambrian socle. It was the advancement of modern geological prospecting with the aid of airplanes, jeeps, and radio that finally
enabled the thorough exploration of the sediment basin.[29] The 1948
drilling of Zelfana (see Chapter 7) was a result of these activities.

On October 16, 1952, the government in Algiers granted the first exploration permits to the French companies Compagnie Francaise de Pétrole (CFP) and Société Nationale de Recherche et d'Exploitation des Pétroles en Algérie (S.N. REPAL).[30] In 1953 further concessions were given to the Compagnie de Recherches et d'Exploitation du Pétrole au Sahara (CREPS, 51 percent ERAP with 35 percent Shell participation) and the Compagnie des Pétroles d'Algérie (CPA, also French with Shell participation). Between 1954 and 1956 concessions were granted to two subsidiaries of the national French Bureau de Recherches de Pétrole (BRP), so that at the end of 1957 almost the entire area (approximately 730,000 square kilometers, or 36.5 percent of the Algerian Sahara) of the Saharan sediment basins, except for the Great Erg, were distributed among oil companies.

The first drilling, by CREPS, in the In Amenas region became productive in January 1956 in Edjéléh, followed by strikes in Zarzaitine, Askaréne, Tin Fouyé, Tiguentourine, and Ohanet in February. Only a short time later, in the spring of 1956, did S.N. REPAL and CFP discover deposits in the Haoudh El Hamra area, particularly those near Hassi Messaoud. In December of the same year the large natural gas field in Hassi R'Mel was tapped by the same associated firms.

Production started in 1958 (a six-inch "baby pipeline" from Hassi Messaoud to Touggourt, thereafter transported by railroad), with a great upswing in production in November 1959 after the completion of the Haoudh el Hamra–Bejaia (Bougie) and the In Amenas–La Skhira (Tunisia) pipeline in September 1960. Verification of these oil deposits also attracted foreign oil companies that profited especially from the redistribution of those concession areas returned by other companies. Nevertheless, the French companies' share of exploration expenses amounted in 1962 to 70 percent of the total (after 90 percent in 1958 and 85 percent in 1959/60). The total investments of all oil companies in the decade 1952–62 were as shown in Table 3.6 (prices reflect the French franc (F) at that time). These expenses were covered by a cash capital of F 2 billion, F 2 billion in loans, and F 3.2 billion from oil sales.

The importance of the companies' own financing already proves the profit possibilities of their operations in the country. Nevertheless, the Algerian deposits cannot be exploited very easily. Taking into consideration the capital investments, the producing costs were figured to average F 18 per ton of oil (1962)[31] including the field–harbor transport. This price was considerably higher than in Libya, for example, and in the Middle East oil-producing countries. There not only the distances to the sea are shorter, but also the expected oil reserves per square kilometer of sediment partially reach 8,000 tons (Middle East), while those in Algeria mounted to a mere 500 to 700 tons (1962).

TABLE 3.6

Investments of the Algerian Oil Industry, 1952–62
(in billions of French francs)

	Amount Invested
Exploration (geology, geophysics, test drillings)	2.75
Development (drillings and material)	2.00
Various expenses (buildings, vehicles, services)	0.75
Transport facilities	1.50
Working capital, reserves	0.20
Total	7.20

Source: "L'Algerie et les hydrocarbures," Annuaire de l'Afrique du Nord 1965 (Paris, 1966), p. 69.

These considerations were summarized in 1965:

Algeria has rather rich oil reserves (1962: approximately 700 million tons), but the producing costs are relatively high because of the geologic and geophysical situation and the distance to the coast (partially made up for by the proximity of the potential customers).

The country is rich in natural gas which can be produced cheaply (1962 reserves: approximately 1,500 billion cubic meters). The connected economic problems concern transport and marketing possibilities.

The independence of Algeria did not de jure change the status of the foreign oil producers. Although the sovereign territory of the new state included what until then had been the French Sahara, Algeria agreed, for the Sahara region, to honor all engagements France had entered into and to guarantee all existing mining rights and concessions.[32] Furthermore, a common proportional board was created that was to deal with questions concerning the utilization of Algeria's natural resources, and to influence the corresponding laws and regulations. Further preference to French companies applying for concessions was planned under the stipulation that no other foreign countries made higher offers. This was in direct opposition to the Program of Tripoli (June 1962). On this occasion the liberation movement FLN stated rather carefully that the nationalization of mining and energy resources should be a long-term goal, which, however, was not further delineated. The French companies expressed their

mistrust particularly through a distinct slowdown in the increasing
rate of extraction while exploration investments (geology, geophysics,
test drillings) almost ceased completely. Resulting from these reac-
tions and from a reduction in the list prices of oil, the Algerian revenues dropped, for instance, from DA 370 million in 1963 to DA 337
million in 1964.

As early as 1963, at the CNUCED meeting in Geneva, the Algerian minister of economics presented a plan that, among other things,
demanded for Algeria: participation of the producer country in the subsequent transformation and valorization of the raw products; a development of industries linked with the production and processing of the raw
materials in the producer country itself, and an organization of the
outlets; and an integration of all industrial activities into the overall
economy of the producer country so that the latter can profit from the
effects of its production as much as possible.[33]

These considerations resulted in the French-Algerian Treaties
signed on July 29, 1965, leading to the foundation of the Association
coopérative (ASCOOP). The Algerian government was represented
by SONATRACH, the French by the Société Pétrolière Francaise en
Algérie (SOPEFAL), which was founded specifically for this purpose.
On that basis, the following Algerian concepts were supposed to be
realized: industrialization of Algeria with the aid of French oil companies; and abrogation of the simple seller position and utilization of
natural resources to accelerate the development of the country.[34]
The demands had resulted from thorough analyzation. With regard
to the Arab oil-producing nations, Ghozali the former and current
president of SONATRACH stated, although certainly simplifying:

> In these regions, in spite of their fabulous wealth, poverty
> is the rule and the level of development is low. Except
> those who claim from the beginning that the Arabs are in-
> capable of development whatsoever, and others who attribute
> the situation to fate, all observers have to acknowledge
> this dramatic paradox: The wealth of the regions is the
> cause for their misery, whereas it should have been the
> source of their prosperity. Everyday the wealth has re-
> tarded the development of a nation (the Arab) which had
> once been the origin of a famous civilization that had shown
> proof of its vitality and fertility. . . . Just as the other de-
> veloping countries, Algeria has inherited an oil industry
> exclusively directed to the outside. The producing and
> transporting activities do not have any impact on the other
> economic sectors; all further activities based on oil take
> place outside the country.[35]

The foundation of ASCOOP had been an attempt to bring both states together as equal partners maintaining their mutual interests and, furthermore, to increase the regional benefits of the oil industry beyond the private interests of individual companies. Obviously, the Algerian requests were not fulfilled.

On February 24, 1971, after months of unsuccessful negotiations, the Algerian government decided to seize control of all of the oil fields and their exploitations by taking a 51 percent share in all the foreign companies operating in the country. The production of natural gas and all means of transport (pipelines) were completely nationalized. According to an official Algerian statement, Algeria, in this distinct and irrevocable manner, expressed its desire to control an essential sector of the economy and to use it solely for the benefit of the Algerian nation.[36] All activities in the oil sector were delegated to the national SONATRACH enterprise. In 1972, 50.8 million tons of oil were produced, of which 39.2 million or 77 percent were by SONATRACH in contrast to the 30 percent before the nationalization. Of all of the concession areas for oil and natural gas, 800,000 square kilometers or 98 percent belonged to SONATRACH in 1971. SONATRACH was also associated as operator of the foreign companies credited with an additional 15,000 square kilometers.

Even before February 24, 1971, SONATRACH had exerted more and more influence, and its participation in the various activities developed as shown in Table 3.7.

These nationalizations did not mean a stop in cooperation with foreign companies; the cooperation was founded on a new juridical base. In June 1971, CFP-Total as the first French company made peace with the Algerian government and accepted a $61 million remuneration. At the same time, CFP started a 49 percent participation in the Algerian ALREP (together with SONATRACH) and signed a contract to invest $100 million between 1971 and 1975.[37] This contract is being carried on until 1980 when all interest in CFP will pass on to SONATRACH.

Further partnership contracts were signed with Sun Oil (April 1973), Hispanoil, Copex (Poland), Petrobras (Brasil), Deminex (Germany), and the French ERAP (January 1974, ended December 1975).[38] In these contracts, SONATRACH generally holds 51 percent. The foreign companies must assume the expenses of exploration and all risks involved. Additional joint enterprises were conducted with Getty Oil and Amoco in 1975.[39] They contain a concession area of 95,500 square kilometers. The companies are obliged to invest a total of $300 million until approximately 1980. Parts of that sum have to be paid directly to SONATRACH for services already performed.

TABLE 3.7

SONATRACH Participation in the Activities of the Algerian Oil
Industry, 1966–73
(in percent)

	1966	1967	1968	1969	January 1970	March 1971	January 1972	1973
Operator on concessions	12.0	21.0	51.0	65.0	92.0	100.0	100.0	100.0
Oil production	11.5	11.8	13.7	16.2	21.2	62.8	76.8	76.3
Gas reserves	18.0	19.5	19.5	23.5	29.0	100.0	100.0	100.0
Oil/natural gas transport	38.0	38.0	39.0	40.0	50.5	98.0	100.0	100.0
Petroleum distribution network	0	48.6	100.0	100.0	100.0	100.0	100.0	100.0

Sources: La Razette (Algiers), no. 18 (February 1972); and
Annuaire Statistique de l'Algérie, 1974 (Algiers, 1975), p. 143.

Current Situation and Economic Importance

Although this study is predominantly concerned with regional
effects, an overall view is indispensable to convey information and
understanding of the Algerian situation. The most important instru-
ment of the present Algerian oil politics is SONATRACH, founded on
December 23, 1963, which by decree 66–292 of September 22, 1966,
extended its tasks to

1. performance in all activities of exploration, technical and com-
    mercial exploitation of humid and gaseous hydrocarbons and re-
    lated substances;
2. construction, technical and commercial service of all sea and
    land hydrocarbons transportation means;
3. processing of hydrocarbons in Algeria and foreign countries;
4. construction, or purchase and renting, of facilities for the indus-
    trial treatment of solid, humid, or gaseous hydrocarbons, in par-
    ticular the development of a petrochemical and hydrocarbon in-
    dustry;
5. distribution and sale of hydrocarbons and their derivatives in Al-
    geria and foreign countries;

6. administration in own name of the title deeds in the above-mentioned sectors, with Algeria holding title to these deeds;
7. participation in industrial construction and realty activities directly or indirectly connected with the above-mentioned sectors.[40]

In Algeria, SONATRACH has a de jure or quasi-monopoly, particularly in activities 1, 2, 3, and 5. These far-reaching assignments clearly show the importance of the oil/natural gas sector to the entire Algerian economy.

Actually, during the Three-Year Plan 1967-69, a total of DA 2.71 billion, that is, 50 percent of total public investments, was invested in hydrocarbons and the chemical industry. During the Four-Year Plan 1970-73, this amounted to DA 4.573 billion or 36 percent of total industrial investments. In 1972, 15,000 people and in 1973, approximately 23,000 people were employed by SONATRACH, of which 99 percent were Algerians. Algerian specialists, whose number rose from 470 in 1971/72 to 1,950 in 1973, are being trained in the Algerian Petroleum Institute (418 students in 1971/72) and the Institut National des Hydrocarbures (2,093 students). With an annual output of approximately 50 million tons, SONATRACH declares itself to be the world's tenth largest oil producer. The extent of its economic activities shows, with a certain stagnation in output, a steady increase in oil revenues (see Table 3.8).

Although oil revenues do not attain the level of some other Arab states (see Chapter 1), the monetary influence of the oil industry on the financial structures of the state is evident. State incomes from oil sales (including operation surpluses and taxes paid by SONATRACH) developed as shown in Table 3.9. Whereas in 1974 oil revenues were supposed to cover the total investment budget of DA 6.5 billion (incidentally, the revenues rose to DA 11 billion), they have now surpassed the investment demand (DA 8.7 billion in 1975 as well as in 1976) and are partially used within the general budget. Following the high foreign currency investments in industry, foreign debts rose to about 35 percent of the 1976 GNP of about DA 46 billion.[41]

As a result of a certain decline in oil prices after the peak of $13.11 per barrel in 1974, revenue estimates from oil of DA 18 billion per annum did not materialize. In 1976 SONATRACH aimed at an average price of $13.36 per barrel, a figure not reached by the actual 1976 range of $12.70 to $12.90 for light and Zarzaitine crudes.[42] But, in the future, additional incomes will be derived from the sale of petrochemical products (see Chapter 10) and especially from natural gas exports (further details in Chapter 12). Gas reserves were estimated, in 1975, at 3.570 billion cubic meters. Sales contracts covering the export of 54 billion cubic meters annually to Europe (22 billion) and to the United States (32.2 billion) already had been signed

TABLE 3.8

Activities of the Algerian Oil Industry: Extent of Exploration,
Production, and Profits, 1952–76

Year	Geo-physics, Team/ Months	Exploration Drilling, 1,000 Meters	Production Drilling, 1,000 Meters	Oil Ex-ports in Million Tons	Duties to the State in Million DA
1952	82	1	0	—	—
1953	113	1	0	—	—
1954	116	31	0	—	—
1955	197	38	0	—	—
1956	225	59	8	—	—
1957	231	75	50	—	—
1958	248	79	149	0.4	—
1959	282	87	225	1.2	—
1960	269	125	254	8.6	—
1961	338	142	271	15.8	—
1962	242	147	194	20.7	—
1963	195	182	210	23.9	—
1964	128	130	121	26.5	—
1965	41	50	156	26.5	—
1966	45	52	123	33.9	—
1967	116	57	101	39.0	—
1968	116	62	140	43.0	920
1969	108	99	148	47.0	1,050
1970	153	107	247	45.5	1,150
1971	220	69	225	34.9	1,600
1972	224	34	223	46.8	3,200
1973	231	45	198	45.0	4,110
1974	—	—	—	43.7	11,000
1975	—	—	—	37.5	13,000
1976 (esti-mates)	—	—	—	—	15,000

Sources: SONATRACH; Petroleum Press Service 40, no. 11
(1974): 416; 41, no. 7 (1974): 266; and Annuaire statistique de l'Algeria 1974 (Algiers, 1975), p. 144.

TABLE 3.9

Part of Oil Income in Algerian State Revenues, 1971–76
(budget figures)

Year	Total State Revenues (1)	Oil Duties of (1) (2)	Percent of (2) in (1)
1971	7,500	1,600	21.3
1972	8,702	3,200	36.8
1973	10,310	4,110	39.9
1974	14,181	6,500	45.8
1975	21,995	13,000	59.1
1976	24,190	15,000	62.0

Sources: "Loi de Finances et mesures sociales," Moudjahid, January 8, 1974; and G. Pierre, "Les budgets de l'Algérie, du Maroc et de la Tunisie," Maghreb-Machrek, no. 72 (1976), pp. 39–54.

in the fall of 1973.[43] The sales prices demanded by Algeria of up to $1.50 per million BTUs (British Thermal Units) have so far not been accepted by all consumers.[44] If this charge materializes with the onset of transport in the 1980s, it would result in an additional income of about $229.5 million annually. As the exploration and development activities have been neglected in Algeria in the past years, and coupled with the fact that domestic consumption had grown from 3.6 million tons (1973) to 5.1 million tons (1975),[45] the amount of oil actually exported was less than anticipated. It seems, however, that this stagnation fits into Algeria's energy strategy as long as it is compensated by the rising oil prices.

The oil reserves, which were estimated at 962 million tons in 1975, have not increased much since the 1962 estimates of 700 million tons. Although new deposits will probably be found, especially in the Great Erg region, Algeria will certainly never be considered one of the reserve-rich producers. A strategy of investment in an industry must replace oil as the motor of economic development. In fact, recent oil revenues have enabled the state to set up generous planning concepts for the future. For example, the Second Four-Year Plan 1974–77 suggests a DA 110 billion volume of investments, 65 percent of which will be covered by oil revenues.[46] Of this amount, 43.5 percent is to be spent in industry, 15 percent in agriculture, and 14 percent for improvements in the infrastructure. At least 450,000 nonagricultural jobs are supposed to be created along with an annual

8 percent increase in the per capita income. The revenues will result predominantly from petroleum, whose production was projected
to reach 65 million tons per year by 1977, but, in reality, it will only
attain 42 to 46 million tons in 1976.

Oil and Development Strategy

As shown in the preceding chapters, Algeria, since its independence, wanted to free itself from the role of a simple, "passive" raw
material supplier. The continuity of the "oil strategy"[47] subsequent
to the following ideas is astonishing.

As early as 1964 the then president of Algeria demanded that
the oil industry should no longer be isolated from the other sections
of the national economy. He maintained that it should be developed
on the basis of the utilization of the hydrocarbon deposits (mineral oil
and natural gas).[48] A better way must be found, he said, for distributing the direct and indirect effects of industrialization throughout
producing and consuming countries. He also indicated that the natural
resources must be made the starting point for the industrialization of
the country in which they were found. This basic idea has been elaborated in collaboration with prominent economists so that Algeria
now feels justified to act as one of the spokesmen for the raw-material-
producing states of the Third World.[49]

Oil and natural gas can have the following effects on Algeria's
domestic economy:

They can lower the cost of energy (motor fuel, gas, electricity) for
    all demand sectors resulting especially in the attraction of indus-
    tries with a high energy input.
They can serve as primary materials for secondary industries (re-
    fining, plastics, fertilizers, and so on).
The local agriculture can be intensified through use of modern mate-
    rials (fertilizers, chemicals, equipment).
Indirect effects also become effective: cheap energy in the form of
    piped or bottled domestic gas may be supplied to private households,
    thus raising their standard of living and avoiding the destruction
    of natural ligneous vegetation.

But the oil industry based on such purposes must definitely be integrated into the national economy: "Oil and gas as such are not capable of performing such effects. Numerous producing countries have
become increasingly dependent on foreign states. Natural oil and gas,
therefore, must be subject to an efficient plan of action expressing
the will of the producing country."[50]

From this follows inevitably a need for national control of the natural resources with the goal being the optimum utilization of their direct and indirect potential.[51] This means that the external economic benefit would derive from the insistence on the highest possible prices for crude oil, and the domestic economic benefit should ensue from the use of the inflowing capital and the regional advantages, whether man-made or natural.

According to the Algerian concept,[52] the only chance for the developing oil-producing countries is to avoid exportation of the natural resources without making a profit for their domestic economies and to take advantage of their locational benefits before the natural resources are exhausted. The all-important date in this regard is 1995, because by then, barring very improbable significant new finds, Algeria's oil reserves are likely to have run out.[53] Algeria has its last chance today to avoid the waste of its riches of nature without having benefited the national economy.

Following the nationalization of the oil deposits and their production carried out in one Arab state after another since 1971, the next step should be an industrialization based on this raw material itself. As recently formulated by the OPEC states, "a substantial proporation of the oil processing [petrochemicals, fertilizers] should be undertaken in the OPEC countries with the [technological] assistance of the industrialized states to which a large part of the output should be exported. At the same time the industrial countries must do away with the discriminatory measures against industrial products from the developing countries."[54] The translocation of oil-based industries to the oil-producing countries may even result in advantages (cost savings) for the industrialized consumer countries.[55]

In fact, mechanisms of the world market do not favor such an industrial development. Industrial activities promoted from abroad were either oriented toward the primary transformation of raw materials or were labor intensive and based on cheap labor costs, as a "prolonged work bench." But it is the aim of Algeria to establish an economy of self-sustained growth with its own potential for development,[56] which is powered from the outset by an industry, including a heavy and capital goods industry. Amin stated as early as 1966 that an accelerated economic growth of the Maghreb countries could not be based on further integration into the world market.[57] The demand of the latter is too elastic for agricultural products, whereas the goods of the new industries cannot compete with similar products from developed countries of higher productivity. This led to the notions of an integrated development on a purely national level, and of the "fight against the international labor division,"[58] seen only as a vehicle of "international neocolonialism."

In fact, such a division principally favors the already-industrialized countries as they keep their monopoly on technological prog-

ress and are always at its head. They even profit from industrialization in developing countries, as they are producers of needed equipment.[59] Thus, industrialization has brought financial problems again:
The new industrial equipment needs foreign currency incomes to
cover expenses for investment and current costs, and these incomes
can be realized only by crude oil and gas sales. At falling world
prices, Algeria would thus be forced to increase its production again,
as higher oil prices are unobtainable because of the fall in energy consumption.[60]

Among the problems facing the Algerian concept of integrated
industrialization, two are external: (1) the lack of "technological capital," which, in Algerian opinion, is kept back in the form of "industrial secrets";[61] and (2) the size of the national market, which is certainly too small to constitute the base of an economy with self-sustained growth. Such growth should host conplementary production
factors and sites as well as an additional demand.[62] The creation of
a bigger Maghreb Common Market is therefore postulated by economists,[63] but it seems unlikely to be achieved in the near future.

It follows from what has been said that the Algerian oil strategy,
and by analogy the energy policies of the other populous oil-producing
countries, will have the following aims in relation to the world energy
market:

Exploration of all potential oil and natural gas deposits so as to gain
    a general view of the extent of these resources and the possibility
    of exercising control over them.
Regulation of oil and natural gas production, not only in terms of de-
    mand from the oil-importing countries, but also in terms of the
    fulfillment of local requirements—hence conservation of the reserves
    and stabilization of oil earnings. Such conservation is very impor-
    tant because oil and natural gas are not only a source of energy but
    also a raw material for the production of innumerable important
    products, especially fertilizers.
Preservation of the purchasing power of the oil revenues.[64]

An indication is the fact that the Four-Year Plan 1974-77 anticipated an annual production of 65 million tons of crude oil, although an
official publication in 1971 spoke of soon expanding the production to
90 million tons.[65] Also, during a dispute over the nationalization,
SONATRACH proposed in May 1971 the possibility of easily raising
the production of the Hassi Messaoud field from the present 20 million tons per year to 40 million or even to 50 million. In contrast to
all of these predictions, the production has remained generally on the
same level. Exports even declined because of a rise in domestic
consumption from 1.8 million tons (1965), to 3.6 million (1970), and
to 5.1 million in 1975.

Today, Algeria is not counted among the poor countries of the world. The enormous revenues from the oil industry and other natural, infrastructural, and social prerequisites allow it to follow a rather independent domestic and foreign policy and even to be an interlocutor in disputes over the trade policy between industrial and raw-material-producing countries. The country's entire oil strategy is based on an evaluation of the realities of the domestic economy. This will be presented in the following chapters and will provide an insight into the bases for the present and future decisions of oil-producing countries. These decisions are no longer made at the headquarters of the international oil concerns but in the producer countries in which the regional effects of the mining industry—directly or indirectly—dominate.

NOTES

1. World Bank Atlas, 1975.
2. Very useful information on economic data and the problem of industrial cooperation is available from Abderrahmane Robana, The Prospects for an Economic Community in North Africa (New York: Praeger, 1973); for the latter question, see also Ghali Boutros, "The League of Arab States," in El Ayouti, ed., The Organization of African Unity (New York: Praeger, 1975), pp. 47–61. For a general "classical" survey, see William Zartmann, Government and Politics in Northern Africa (London, 1964); and, more recently, W. Zartmann, "The Southwest Shore," International Journal (Toronto) 21, no. 4 (1972): 593–605.
3. From "Rôle et interventions des services de la Direction des Domaines et de l'Organisation foncières dans l'application des mesures de Révolution agrairen, La Revue financiere (Algiers), no. 2 (1971), pp. 167–76. Similar figures are advanced recently by B. Cherbal, "L'agriculture algérienne—organiser la production," Moudjahid, September 9, 1975. For a general survey of Maghreb agrarian problems, see Horst Mensching, "Le milieu naturel du Maghreb: Questions et limites de la mise en valeur du potentiel agraire," in W. K. Ruf et al., eds., Introduction à l'Afrique du Nord contemporaine (Paris, 1975), pp. 49–61; and Elias H. Tuma, "Population, Food and Agriculture in the Arab Countries," Middle East Journal 28, no. 4 (1974): 381–95.
4. Figures from Plan quadriennal 1974–77; see also Dj. Sari, "L'évolution de l'emploi en Algérie," Maghreb-Machrek, no. 69 (1975), pp. 42–50. B. Cherbal, "La recherche d'un équilibre géo-économique," Moudjahid, May 17, 1975, indicates a figure of 586,843 farmers in the private sector. The positive impact of the agrarian revolution has been rather limited, as only 54,000 formerly landless

peasants have, until 1975, been beneficiaries of land distribution (see Sari, "L'evolution de l'emploi en Algerie," p. 47).

5. See "La situation de l'agriculture algérienne en 1967," Statistique agricole, no. 7 (1968), p. 19. Figures forwarded by the Ministry of Agriculture and Agrarian Reform following a 1972/73 survey give a still more desperate image: They enumerate a total of 899,545 agricultural exploitations, of which 168,683 or 18.7 percent have no land and 161,667 or 18.0 percent have less than 1 hectare. They possess 5,207,611 hectares of agriculturally useful land or 5.8 hectares per exploitation. And in each farm there is an average of 4.7 adult persons and of 2.2 agriculturally active persons. See Annuaire statistique de l'Algérie 1974 (Algiers, 1975), pp. 89 ff.

6. Following La situation de l'emploi et des salaires 1970 (Algiers, 1970), p. 44.

7. Calculations according to Tableaux de l'économie algérienne 1971, pp. 120-21; and "La situation de l'agriculture algérienne," p. 19.

8. Personal calculations based on purchase prices and wholesale prices, respectively, in Algeria. Algérie—Situation économique 1972-73, pp. 25-72 and 317-19; partially also extrapolations from Annuaire statistique de l'Algérie 1970 (Algiers, 1971), pp. 120 and 220-23. Viratelle, however, forwards different data without naming any sources. According to his figures, the private sector produced about 55 percent of the production and received 38 percent of the agricultural incomes. See G. Viratelle, L'Algerie algerienne, Coll. Developpement et Civilisations (Paris, 1973). Certainly, the self-supporting function of the private agricultural sector is not taken into consideration. Production figures did not change much until 1973 (most recent figures available), but the weighted price index (food prices at consumer's level in Algiers, base 1969 = 100) rose from 109.4 in 1971 to 120.4 in 1973. See Annuaire statistique de l'Algérie 1974, p. 219.

9. Calculations from Situation économique, 1972-73, pp. 30-31.

10. Monthly Bulletin of Agricultural Economics and Statistics 23, no. 3 (1974): 13.

11. See Konrad Schliephake, "Changing the Traditional Sector of Algeria's Agriculture," Land Reform, no. 1 (1973), pp. 19-28.

12. See Jean Le Coz, "Mutations rurales au Maghreb: Du dualisme agraire a l aménagement de l'éspace," in W. K. Ruf et al., eds., Introduction à l'Afrique du Nord contemporaine (Paris, 1975), pp. 63-80.

13. See, on cereals scarcity in the world and the problems of an emerging producer's cartel, A. Belhimer, "Le blé—une pénurie entretenue," Moudjahid économique, February 24, 1976.

14. From Statistiques financieres (Tunis), nos. 36-38 (November 1975), p. 53.

15. See Sadok, "Budget 1975," El Djeich (Algiers), no. 141 (1975), p. 18; and "Le budget 1976 de l'Algérie," Marchés tropicaux, no. 1576 (1976), pp. 185-89.

16. See Salem A. Hajjaji, "Agricultural Development and Land Settlement in the Kufra Region of Libya, Land Reform, no. 1/2 (1974), pp. 68-88; and Konrad Schliephake, Libya—Economic and Social Structures and Development (in German), Arbeiten aus dem Institut für Afrika-Kunde, vol. 3 (Hamburg, 1976).

17. See Adolf Arnold, "Die Industrialisierung in Tunesien und Algerien," Geographische Rundschau (Braunschweig) 23 (1971): 306-16.

18. As most of the statistical figures still refer to the former units, the reshaping was not considered further. For details, see Moudjahid, July 3, 1974, with a map of the new boundaries.

19. See Boualem Bakour, "Les programmes spéciaux ou la lutte contre les disparités et deséquilibres régionaux," Nouvelles économiques (Algiers), nos. 122-23 (November 1 and 15, 1972), pp. 6-7; figures on definite spending from B. Cherbal, "La recherche d'un équilibre géoéconomique," Moudjahid, May 17, 1975.

20. Citation in "Front contre le deséquilibre régional," Nouvelles économiques (Algiers: CCIA), no. 135 (August 1, 1973), p. 1.

21. This number is calculated from the difference between the 1966 census and the population estimate on April 4, 1971. In particular, it also corresponds mostly with the birth/death statistics (see Annuaire statistique de l'Algérie 1970 [Algiers, 1971], p. 21) and clearly differs from the figure of 3.2 percent and the estimate of 3 percent that often can be found in publications. See also Keith Sutton, "Algeria: Changes in Population Distribution, 1954-66," in Populations of the Middle East and North Africa, ed. J. I. Clarke and W. B. Fischer (New York, 1972), pp. 373-403.

22. According to Dj. Sari, "Problèmes démographiques algériens," Maghreb-Machrek, no. 63 (1974), pp. 32-42; and Annuaire statistique de l'Algérie 1974, p. 23.

23. See G. Negadi, D. Tabutin, and J. Vallin, "Situation démographique de l'Algérie," Dossiers documentaires, nos. 19-20 (Algiers, 1972), pp. 13-31.

24. Viratelle, L'Algérie algérienne, pp. 217 and 220.

25. A. Tiano, "Human Resources Investment and Employment Policy in the Maghreb," Employment in Africa (Geneva, 1973), pp. 151-75.

26. J. P. Sereni, "Prélude au rendez-vous 1980," Jeune Afrique, no. 706 (July 20, 1974), pp. 34-35.

27. G. Boutaleb, "Priorité à l'équipement et à l'investissement," Révolution africaine (Algiers), no. 568 (January 10, 1975).

28. J. P. Hullot, "Le pétrole en Algérie," Revue des Etudes politiques africaines, no. 47 (April 1969), pp. 18-37.

29. E. Klitzsch, "Über den Grundwasserhaushalt der Sahara," Afrika Spektrum 3 (1967): 25-37.

30. Here and for the following, see, "L'Algérie et les hydrocarbures," Annuaire de l'Afrique du Nord 1965 (Paris, 1966), pp. 63-100;

O. Alber, "Erdöl und Erdgas in der Sahara," Afrika Spectrum, no. 3 (1967), pp. 38–47; and Henri Madelin, Pétrole et politique en Méditerranée occidentale, vol. 188 (Paris: Cahiers de la Fondation nationale des Sciences politiques, 1973), pp. 144–87.

31. "L'Algérie et les hydrocarbures," p. 75.

32. G. Destanne de Bernis, "Les problèmes pétroliers algériens," Revue algérienne des Sciences juridiques économiques et politiques 10, no. 4 (1973): 717–58.

33. Ibid., p. 723.

34. See A. Ghozali, "L'association coopérative algéro-francaise: ses buts, son bilan, ses perspectives," Revue algérienne des Sciences juridiques économiques et politiques, no. 1 (1969), pp. 165–72.

35. Ibid., pp. 165, 168.

36. "Le pari algérien," Jeune Afrique, no. 633 (February 24, 1973), pp. 22–30 (various articles).

37. "CFP Makes a New Attempt," Petroleum Press Service (London) 40, no. 7 (1973): 265–66.

38. "Relance de recherche au Sahara," Pétrole Informations, no. 1300 (January 25/31, 1974), p. 20; and "SONATRACH et Deminex signent un protocole pour l'exploration et la production de pétrole," Pétrole et Gaz arabes (Beirut), no. 114 (December 16, 1973).

39. Industries et Travaux d'Outremer, April 1976, p. 265.

40. According to various official presentations of SONATRACH. Also numerous explaining publications, such as "WONATRACH's Oil Empire," Petroleum Press Service (London 39, no. 2 (1972): 52–54; Hassan, Zenati, "La SONATRACH, première compagnie pétrolière du Tiers monde," Jeune Afrique, no. 539 (May 4, 1971), pp. 28–32; and "SONATRACH—première société pétrolière nationale du Tiers monde," Europe Outremer, no. 533/534 (1974), pp. 41–42.

41. L'amélioration, en 1972, de la conjoncture économique algérienne," Marchés tropicaux, no. 1452 (September 7, 1973), pp. 2673–74, with a list of the loans taken out in 1972; and "Le budget 1976 de l'Algérie," Marchés tropicaux, no. 1576 (1976), pp. 185–89.

42. Petroleum Economist, March 1976, p. 113.

43. "SONATRACH Changes Terms of Delivery," Petroleum Press Service 40, no. 10 (1973): 365–67, a view over all projects; for more details see Chapter 6.

44. Petroleum Economist 41, no. 8 (1974): 309–10; also Pétrol et gaz arabes, no. 116 (January 16, 1974), p. 15.

45. Figures published by ESSO, 1976.

46. See, among others, "Le IIème Plan quadriennal," Algérie Informations, no. 4 (Paris, July 15, 1974), p. 19; "Le Plan: très bonnes perspectives au début du deuxième plan," Europe Outremer, no. 533/534 (1974), pp. 6–11; "Le deuxième plan 1974-1977: use gigantesque bataille pour la production," Europe Outremer, no. 542 (1975), pp. 29–34. Preenergy-crisis drafts had foreseen a total investment volume of only DA 52 billion.

47. Out of the large amount of special literature in French, see, for instance, G. Destanne de Bernis, "Les industries industrialisantes et les options algériennes," Revue Tiers-Monde, no. 47 (1971), pp. 545-63; Jean-Pierre Sereni, "La politique algérienne des hydrocarbures," Maghreb, no. 45 (1971), pp. 31-49; in a geographical approach also Djilali Sari, "La récupération et la valorisation des hydrocarbures par l'Etat algérien," Annales algériennes de Géographie, Special number (1972), pp. 209-37; in a political approach see, for instance, Madelin, Pétrole et politique en Méditerranée occidentale, pp. 78-83; and, rather recent and comprehensive, Bruno Etienne, "La place des matières premières dans la politique extérieure de l'Algérie," Annuaire de l'Afrique du Nord 1974 (Paris: CNRS, 1975), pp. 71-92.

48. A. Benbella, "La politique pétrolière algérienne," Revue algérienne des Sciences juridiques économiques et politiques, no. 4 (1964), pp. 83-103; see also "L'Algérie et les hydrocarbures," Annuaire de l'Afrique du Nord 1965 (Paris: CNRS, 1966), pp. 63-100.

49. Azzedine Mabrouki, "L'O.P.E.C. et les Tiers-Monde—un même combat," El Djeich (Algiers), no. 143 (April 1975), pp. 29-34.

50. A. Ghozali and G. Destanne de Bernis, "Les hydrocarbures et l'industrialisation de l'Algérie," Revue algérienne des Sciences juridiques économiques et politiques, no. 1 (1969), pp. 253-94; and the Memorandum presented by Algeria to the Conference of OPEC Sovereigns and Heads of State (Algiers, March 1975).

51. Here the problems of nationalizations cannot be discussed, but UN Resolution 626 (VII) from December 21, 1952, should be kept in mind, which stresses the continuous sovereignty of every country on natural resources and potentials, and the right to deal freely with them. See, for the concrete case, G. Vlachos, "Le régime juridique des hydrocarbures en Algérie," Revue juridique et politique Indépendance et Coopération (Paris) 28, no. 1 (1974): 103-28; A. C. Zoeller, "Algerian Nationalizations: The Legal Issues," Journal of World Trade Law 6, no. 1 (1972): 33-57; J. Basso and J. Touscoz, "Les stratégies maghrébines pour l'exploitation et l'exportation des matières premières: Quelques problèmes de droit international public," Annuaire de l'Afrique du Nord 1974 (Paris: CNRS, 1975), pp. 173-205; also the general and very critical studies of R. R. Odell, Oil and World Power: A Geographical Interpretation (Harmondsworth, 1970); and M. Tanzer, The Political Economy of International Oil and Underdeveloped Countries (Boston, 1970).

52. See Le pétrole, les matières de base et le développement—Mémoire présentée par l'Algérie à l'occasion de la session extraordinaire de l'Assemblée Générale des Nations unies (Algiers, April 1974), pp. 53-54.

53. "Algeria—Preparing for the Take-Off," Financial Times, April 24, 1975.

54. "Solemn Declaration Adopted by the First Conference of OPEC Sovereigns and Heads of State on March 6, 1975, in Algiers," Memorandum Presented by Algeria, pp. 41–48.

55. See Farid Akhtarekhavari, Die Ölpreispolitik der OPEC-Länder [The oil price policy of the OPEC countries], Probleme der Weltwirtschaft, Diskussionsbeiträge, no. 2 (Munich, 1975).

56. The Industrial Revolution (Algiers: Ministry of Information, 1975), p. 19.

57. Samir Amin, L'économie du Maghreb (Paris, 1966), p. 192.

58. "Les fondements de la planification algérienne," El Djeich (Algiers), no. 138 (November 1974), pp. 75–79.

59. See A. Ghezali, "L'impérialisme de l'an 2000 en construction," Moudjahid, September 9, 1975; Issam El Zaim, "Décolonisation économique, développement et multinationales," Moudjahid, October 14, 1975.

60. "Raising State Revenues Despite Falling Exports," Petroleum Economist, March 1975, pp. 84–86.

61. Mémoire présenté par l'Algérie, pp. 104–05.

62. See Amin, L'économie du Maghreb.

63. See Robana, The Prospects for an Economic Community.

64. Memorandum Presented by Algeria, p. 70.

65. M. Aliev, et al., Geological Structures and Estimations of Oil and Gas in the Sahara in Algeria (Algiers: SONATRACH, 1971), p. 239.

II

REGIONAL
EFFECTS OF OIL
EXPLORATION AND
PRODUCTION

4

EFFECTS OF EXPLORATION

TYPES OF OIL ACTIVITIES

As explained in Chapter 1, this study will deal predominantly with the primary and regional effects of the oil industry. To examine these effects we have to make these divisions: prospecting of oil, production of oil, transport and storage of oil, processing of oil (refineries, petrochemical industries). No research has been done yet, at least not on North Africa, concerning the questions posed here. However, their answers are of utmost importance to the producer countries in their regional planning. They are of equal importance to the foreign oil companies operating in these lands. Because of economic and political reasons, the companies cannot afford to neglect the side effects of their oil-producing activities. The governments of the developing countries pay particular attention to developmental and welfare effects coupled with raw material extraction and processing. The following chapters will present case studies for each division of oil activities and try, if possible, to derive general tendencies.

LABOR DEMAND

There is almost no information on the effects of exploration activities (search and development of oil deposits) upon the regional economy. However, there is a demand for labor becoming evident in those areas without further job offerings in the modern industrial sector. During the peak of the exploration activities in the Fezzan (Libya), the city of Sebha experienced a true "oil fever" with a rapid increase in population and a parallel building boom. Unskilled laborers who came from the traditional sectors went through a change in mentality, as could be observed in Libya.[1]

These changes in motivation and social behavior are also observed in the Algerian Sahara, although not within oil activities. For
the French nuclear center in In Ekker (Hoggar), operating from 1959
until 1966, enormous construction endeavors were undertaken employing an average of 800 to 1,200 workers over periods of several
weeks to several months. The annual labor assignment of 2,000 to
3,000 natives was comprised of small farmers from nearby cultivation
centers (that is, Tazrouk) and some nomads for whom these jobs meant
occasional additional income. These nomads experienced for the first
time a regular monetary allowance connected with the advantages of
being stationary. After the shut down of the nuclear center, the small
farmers went back to their fields and some of the nomads returned to
their migratory life style. Their main aim had been to escape their
misery and look for a regularly paying job. The activities at the In
Ekker camp had been an intermediate station between nomadism and
sedentary life.[2]

Aside from these changes in behavior, of which only tendencies
can be shown, the question of quantitative effects must be raised.
Neither within the oil companies nor in official statistics for any of
the Arab countries can any information be obtained about the work force
employed in exploration activities, nor wages paid. Information from
Libya stated that 30 to 60 persons worked at each drilling installation
and approximately 80 in each seismic group. These figures could be
roughly verified by personal surveys in Tunisia and Algeria so that
the calculations in Tables 4.1 and 4.2 are based on 80 men per exploration team and 50 men per derrick team.

These numbers show the very small regional importance of such
activities, which become even more distinct by taking into account
that some workers of the drilling parties are foreigners (for example,
at the exploration drilling Keskessi/El Borma: 9 foreigners among 51
workers; exploration drilling Hassi R'Mel: 5 among 77 workers). In
addition, only some of the workers come from the region itself (that
is, Keskessi: 60 percent; Hassi R'Mel: 57 percent). In particular,
the higher-paid workers come from the bigger cities, in the north.

MONETARY FLOWS

A calculation of the monetary effects according to these facts
results in the following estimates (1971) for the Wilaya Oases where
the Algerian exploration was almost exclusively carried out·

Yearly average of the work force de- mand (1971)	2,490
Minus 10 percent of the foreigners	2,241

TABLE 4.1

Oil Exploration in Tunisia: Size and Demand for Manpower,
1954–70

Year	Geophysics Team/ Months	Meters Drilled in 1,000[a]	Drilling Team/ Months[b]	Monthly Average of Workers[c]
1954	131.8	13.6	33.9	1,019
1955	74.2	18.0	41.5	667
1956	60.8	9.9	21.2	493
1957	22.8	9.8	19.3	232
1958	58.0	9.8	18.2	463
1959	28.5	6.4	11.6	283
1960	53.5	0	0	357
1961	74.5	3.7	5.8	521
1962	88.2	21.9	20.1	672
1963	61.8	7.3	8.6	448
1964	39.0	26.7	31.5	391
1965	81.0	47.0	55.3	770
1966	104.0	27.9	32.9	830
1967	78.0	58.4	68.7	806
1968	41.2	40.9	48.1	475
1969	57.1	26.4	31.0	510
1970	34.5	18.0	25.7	337

[a]Exploration, production and development drillings.
[b]Calculated from the drilling performance during the drilling period.
[c]Our calculations are based on the assumption that each geophysical team consisted of 80 men, each drilling team of 50 men.

Source: In part personal calculations, in part from Tunisia— The Development of the Petroleum Industry, E/CN. 14/EP/58, September 24, 1973 (New York: UN Economic and Social Council, 1973), pp. 40, 48 ff., and 59.

| Minus 30 percent of the workers from Northern Algeria | 1,570 |
| Average wage DA 800 per month (lower than the SONATRACH wages, frequent contractors) results in a yearly total of | DA 15,072,000 |

This means that in comparison with the total income of DA 710.9
million (according to our calculations in Chapter 6), which hypothetically is available for the Wilaya Oases, only 2.1 percent is achieved
by wages of workers employed at the exploration activities (including

TABLE 4.2

Oil Exploration in Algeria: Size and Demand for Manpower, 1954–73

Year	Geophysics Team/ Months	Meters Drilled in 1,000[a]	Drilling Team/ Months[b]	Monthly Average of Workers[c]
1954	165.8	67.3	152.4	1,750
1955	204.1	59.8	141.6	1,951
1956	225.2	66.9	176.4	2,236
1957	228.3	121.4	255.6	2,587
1958	248.7	228.3	399.6	3,323
1959	281.3	312.6	438.0	3,700
1960	269.0	379.7	428.4	3,578
1961	337.3	412.6	402.0	3,924
1962	241.9	340.8	402.0	3,288
1963	200.5	391.4	390.0	2,962
1964	128.0	252.2	267.6	1,968
1965	41.2	204.6	172.8	995
1966	47.3	175.3	141.6	905
1967	99.7	153.1	152.4	1,300
1968	115.5	201.7	159.0	1,433
1969	108.3	247.8	211.5	1,603
1970	153.3	354.0	—	—
1971	220.3	294.0	245.1	2,490
1972	224.0	257.0	214.0[d]	2,385
1973	231.0	243.0	202.0[d]	2,382

[a]Exploration, information, production, development, and in-
jection drillings.
   [b]Number of derricks operating times period of work, in months.
   [c]Our calculations are based on the assumption that each geophysical team consisted of 80 men, each drilling team of 50 men.
   [d]Estimates.

   Sources: Activité de l'industrie pétrolière 1965 (Paris: Direction des Carburants, 1966), pp. 63, 65, and 119; ibid., 1967, pp. 64,
65, 120, and 130; Annuaire statistique de l'Algérie 1972 and 1974
(Algiers, 1973 and 1975), pp. 118 and 145.

production drillings). This is a measurable but not an important amount.

Certain regions, such as Metlili des Chaambas (near Ghardaia) and Djemaa (near Touggourt), however, are preferred suppliers of workers, especially for oil prospecting. The contractors, in particular, recruit their seasonal work force from these places so that during certain periods even a lack of agricultural workers can be felt.[3]

With respect to the regional effects, the anticipated and partly realized high expenses of the oil companies should not be misleading. In Tunisia, for instance, contracts with the oil companies are made under the assumption that 70 percent of the companies' expenses are spent in foreign currencies, that is, in foreign countries, and the remaining 30 percent in TD; for off-shore undertakings, this ratio is 85 percent to 15 percent. Of the companies' total exploration expenses of TD 100 million (1962-72—TD 28 million in 1972 alone) no more than TD 30 million remained in Tunisia.[4] Even from this amount, no more than TD 8.5 million might have been spent as genuine wages. The fact that the oil companies operating in Tunisia were obligated to invest at least TD 47.4 million in exploration activities between 1973 and 1976, with additional investments of TD 16.7 million planned,[5] should be seen under this aspect.

The financial activities of oil companies in Algeria have to be seen from a similar standpoint. In 1973 alone, six foreign oil companies signed exploration contracts with SONATRACH for regions in which oil is suspected—amounting to a total of 95,500 square kilometers. They were obliged to spend $300 million over a two- to five-year period.[6] These activities were partly profitable to the entire country, but hardly to the regions themselves. Similar to its installations, which last only a few months on one spot, exploration is the most transitory and fugitive of all oil activities.

NOTES

1. F. C. Thomas, "The Libyan Oil Worker," The Middle East Journal 15 (1961): 264-76.

2. Raymond Josse, "Problèmes de mise en valeur du Hoggar et de la croissance urbaine à Tamanrasset," Cahiers d'Outremer (Bordeaux), no. 95 (1971), pp. 245-93.

3. B. Abdiche, "Les palmiers de l'espoir," Moudjahid, February 2, 1974.

4. According to information from the Tunisian Ministry for Economic Affairs.

5. IVe Plan de developpement 1973-1976 (Tunis, 1973), p. 252.

6. "Relance de recherche au Sahara," Pétrole Informations, no. 1300 (January 25 and 31, 1974), p. 20; "SONATRACH et Deminex

signent un protocole pour l'exploration et la production de pétrole,"
Pétrole et gaz arabes, no. 114 (December 16, 1973), pp. 7-8; numerous articles in Marchés tropicaux, for example, on June 8, 1973,
p. 1565; June 15, 1973, p. 1627; January 25, 1974.

5

**EFFECTS OF OIL
PRODUCTION IN
TUNISIA**

Contrary to the indirect, monetary effects, direct effects of oil production are rather insignificant on the national level. However, our microregional studies in Chapters 5 and 6 show a certain importance in given regions, and effects are summarized and generalized in Chapter 7. The variety of natural regions in which oil-production locations in Tunisia are found shows that the regional effects depend only partly on the natural situation. To a much greater extent, they depend on the technologic and economic principles and managerial guidelines within the oil economy.

EL BORMA

El Borma is the biggest Tunisian oil field. Its annual production had been approximately 3.5 million tons until 1975, when it dropped to 1.7 million tons. It is under the administration of the Societe Italo-Tunisienne d'Exploitation Petroliere (SITEP) with the Italian AGIP and the State of Tunisia being equal partners. The Tunisian assumption of the Italian oil interests in the Bizerte refinery (see Chapter 9) and in gasoline distribution has not affected Italian participation. The oil field is situated in the unpopulated part of the medenine Governorate directly on the Algerian border that divides the field into approximately equal halves. In June 1972 the two governments made an agreement, among other things, concerning the distribution of the production (two-thirds Tunisia, one-third Algeria) and joint transport to Skhira.[1] Since July 1975, research and development work has been intensified at the site at costs of TD 50 million and it is hoped that production will return to its former level.[2] Although a pipeline from the Algerian part of the field to Hassi Messaoud is planned, it has not yet been con-

structed and Algerian production continues to be shipped via Skhira
(see Chapter 8; for the gas pipeline to Gabes, see Chapter 9).

Labor Market in Medenine Governorate

Although the entire region is economically marginal in relation
to the rest of Tunisia, without any significant trade or industry (aside
from the tourism in Djerba), the population is constantly increasing:
from 237,559 in 1966 to an estimated 253,328 in 1970, or 7 percent.
In 1966 the work force consisted of 132,300 persons. Only 40,577
men (30 percent of the men or 16.7 percent of the total population)
and 1,252 women (0.95 percent or 0.5 percent, respectively) had a
job. An additional 8,203 men (6 percent of the men, 3.5 percent of
the total population, or 20.2 percent of the work force) were unemployed; and 1,469 men (1 percent and 0.5 percent, respectively) had
been unemployed since their school graduation. For the rest, 65,948
(49.8 percent of the women, or 27.7 percent of the total population)
were housewives, 1.7 percent of the total population were students.
An additional 5,087 (4 percent) of the men and 2,988 (2.2 percent) of
the women were retired, disabled, or otherwise unable to work.[3]
In 1966 a total of only 19.7 percent of the population was economically active (Tunisian average: 21.6 percent). The distribution
is shown in Table 5.1. The lack of job opportunities and natural prerequisites for an efficient agriculture turned the Governorate into the
most significant emigrative and passive region of Tunisia.[4] As the
population increase figures are extrapolated, figures since the 1966
population survey cannot be accurately verified. In fact, between 1956

TABLE 5.1

People Economically Active in Medenine According to Sectors, 1966

	Agri-culture	Pro-duction	Ser-vices	Liberal Professions	Total
Medenine Absolute	19,795	9,506	5,952	11,651	46,904
Percent of total	42.2	20.2	12.7	24.8	—
Tunisia					
Percent of total	45.9	19.1	13.2	21.8	—

Source: Les villes en Tunisie (Tunis: Direction de l'Aménagement du Territoire, 1971), p. 464.

and 1966 the entire population of the Medenine Governorate had only grown at the rate of 3.66 percent total or 0.3 percent annually (Tunisian average: 31.7 percent and 2.7 percent, respectively), the urban population of the Governorate (22.3 percent of the population, 1966), however, grew at a 38.9 percent or 3.28 percent rate annually (respective Tunisian average: 47.3 percent and 3.85 percent).[5]

In the individual delegations, the residential population changed as shown in Table 5.2. Only the urban zones and those at the foot of the mountains (Remada) and the delegations with developing tourism and relatively intensive agriculture (Djerba, Zarzis) could report a moderate growth. The population of the mountain regions decreased and the emigration trend in traditional migratory regions (Tataouine, Ghomrassen, Ben Kheddache) continued.

TABLE 5.2

Change in the Residential Population in Medenine According to Delegations, 1956–66

	1956	1966	Percent Total	Percent Annual
Medenine	36,066	39,526	9.6	0.82
Djerba	62,445	68,220	9.3	0.8
Remada	3,795	8,414	122.0	8.1
Ben Kheddache	16,170	15,908	−6.1	−0.4
Ghomrassen	17,253	15,612	−9.5	−0.8
Tataouine	41,925	34,115	−18.7	−0.7
Ben Gardane	26,761	24,957	−6.7	−0.6
Zarzis	29,375	35,567	+21.1	1.89
Total	233,790	242,319	3.66	0.32

Source: Les villes en Tunisie (Tunis: Direction de l'Aménagement du Territoire, 1971), p. 459.

This tendency has persisted since the 1966 census as confirmed by the migration balance. Table 5.3, a comparison of the rural and urban populations of the Medenine and Gabes Governorates with entire Tunisia, displays this. Of all of Tunisia, the Governorate of Medenine and Gabes have the highest percentage of migration losses, Medenine having the highest absolute figures as well. The external migrations are increased by the high degree of mobility of the population: 7 percent moved within the Medenine Governorate. Further, 3 percent

TABLE 5.3

Migration Balances of the South Tunisian Governorates, 1956–71

	Governorate			
	Medenine	Gabes	Sfax	Tunisian Total
1966–71				
Urban population				
Absolute	+3,315	+6,984	+19,361	—
In percent of urban population	+12.2	+18.0	+16.9	+18.2
Rural population				
Absolute	−35,713	−20,223	−11,926	—
In percent of rural population	−40.6	−34.9	−13.7	−18.2
Migration gains/losses				
Absolute	−32,398	−13,239	+7,435	—
In percent of total population yearly	−1.3	−0.6	+0.2	—
1956–66				
Migration gains/losses				
Absolute	−65,882	−24,456	+75,944	—
In percent of total population yearly	−2.7	−1.2	+1.8	—

Sources: P. Signoles, "Migrations intérieures et villes en Tunisie," Cahiers de Tunisie, no. 79/80 (1972), pp. 207–40; and personal calculations.

moved to the three other southern governorates: Gabes (2 percent), Sfax (1 percent), and Gafsa (0.3 percent) (see Table 5.4). These figures are extraordinarily high even for southern governorates.

The internal movements, dependent on the urbanization coefficient (Medenine 16, Sfax 35), resulted in a concentration of the population in urban settlements, above all in the town of Medenine. In addition to the internal migrations, the temporary emigration of workers into foreign countries is of decisive importance. In early 1973 a work force of 11,171 was working (with labor contracts) in foreign countries. This figure breaks down as follows: from Tataouine: 3,211; Remada: 191; Ben Keddache: 720; Ben Gardane: 423; Zarzis: 1,103; Medenine: 316; Ghomrassen: 1,915; Djerba: 3,292. Of these, 9,483 worked in France, 503 in Libya, 95 in West Germany,

TABLE 5.4

Migration of the Southern Tunisian Population, 1966

	Percent Internal Migration	Percent Emigration Outside of the Three Southern Governorates
Sfax	4.0	1.2
Gabes	3.3	3.5
Gafsa	6.0	1.2

Source: M. Seklani, "La mobilité intérieure dans le Sud tunisien," Revue tunisienne des Sciences sociales 7, no. 23 (1970): 163–74.

and 1,090 in other countries. Actually, these figures are probably one-third too low, and thus 15,000 emigrants can be assumed.[6]

The migrations within the governorates and into other regions of Tunisia and foreign lands are therefore the dominating factor in population behavior. The internal migration is brought about by the desire to improve the material living conditions regarding the infrastructure (better education, health care, and residential conditions). The external migration, however, is caused by the absolute lack of job opportunities on the mainland—except for Medenine. Our survey stated that the migrations resulted from economic reasons and not at all from a "longing for faraway lands." Of all those interviewed in El Borma,[7] three (12 percent) had already emigrated, eight (32 percent) were ready to look for jobs in foreign countries out of economic necessity, and four (16 percent) would only move within Tunisia if necessary. Only one person would have gladly gone to a foreign country for subjective reasons. Ties to the homeland have not been affected by so much regional mobility: 20 (80 percent) of the interviewed wanted to spend their retirement where they used to live and where, as a rule, they had built or planned a modern home. Only five (20 percent) wanted to go to Tunis or another governorate capital (Medenine, Gabes) where their families were already living. The roots in their home villages and regions are very deep; the migration within the country and especially to France and Libya is a bitter, economic necessity.

Monetary Flows

In order to study the quantitative effects of the oil industry, the origin and behavior of the oil workers in El Borma have to be examined. On the El Borma base, approximately 150 persons are employed by SITEP, of which only eight are Europeans, the rest being exclusively Tunisian. Sixty-six come from the Medenine Governorate. Also from Medenine are 29 (of 35) workers from the company Sahara Comfort, which is responsible for food supply and services, 23 from the drilling firm SAIPEM, [8] 12 from the national gas company STEG (which supervises the natural gas pipeline from the Garde Nationale. Altogether, a total of 131 persons come from Medenine Governorate. Eight are from Remada Delegation; 14 from Beni Kheddache, Foum Tataouine, and Ghomrassen; 17 from Zarzis; 78 from Medenine and Ben Gardane; and 13 from Djerba. In addition, 60 persons from the Kilani construction firm are also employed in El Borma for the construction and maintenance of roads and tracks. The entire firm of Sahara Comfort is more or less working for the oil company. It organizes the catering in El Borma and the purchase, distribution, and preparation of the food supply, the transport, and the management of the AGIP Motel in Medenine as well. Originally the hotel was planned as a waystation on the route to El Borma. Of the firm's 123 employees in Medenine, 105 also have their residence there.

To summarize, the results of this survey show 267 men in El Borma, that is, 0.6 percent of all men permanently employed in Medenine. Consequently, the quantitative effects of oil production in this region have to be considered as insignificant. Nevertheless, the study must be continued. The monetary flows, which are released within a branch of activities generally regarded as "rich," have to be examined as well as the associated processes that connect oil and the regional economies and other structures.

The following wages were paid to the above-mentioned employees in El Borma: The workers from Medenine are predominantly employed as helpers (manoeuvre) and simple workers, so their monthly pay by SITEP amounts to approximately TD 85. The salaries paid by SAIPEM may be of the same amount. For these two groups, the total wages equal 89 × 85 = TD 7,565 per month. The salaries paid by other companies in El Borma are considerably lower despite bonuses for working in the Sahara. On the basis of our interviews, they might average TD 45 per month, so the paid sum of wages results in 178 × 45 = TD 8,010 per month. In comparison with the wages generally paid, these are still favorable. The situation in other income groups presents the following picture: The salaire minimum agricole garanti (SMAG), which by no means is exceeded in the south but rather evaded, amounts to TD 0.800 per day since the raise in May 1974 (previously TD 0.600 per day.[9] This means that an average agricultural income of TD 16–20 can be assumed.[10]

The salaire minimum interprofessionel garanti (SMIG) amounts to TD 0.832 per day; and according to our experience an average income of TD 20–25 may be assumed. These figures do not appear to be too low when one considers that during the plan period 1973–77 400,000 families (2 million people) will have an annual income of under TD 250 per year (TD 50 per person).[11]

The monthly sum of wages paid in the governorate is calculated in Table 5.5 for the agricultural sectors (the absolute figures have, of course, only relative correctness). In comparison to the total sum of salaries directly or indirectly paid by the oil industry (TD 15,575 per month), these figures have a ratio of 100:1.5. The share of the oil industry has to be characterized as existent but insignificant.

The monetary flows caused by the emigration of workers into foreign countries are far more important. With 15,000 workers abroad, who send home a monthly average to TD 20 to Medenine, their sum amounts to TD 300,000 per month or 28.9 percent of the incomes from local activities (excluding oil). Locally available incomes from oil, emigration, and other activities added, an average of TD 4.75 per month come to each inhabitant. This figure seems realistic in light of the fact that 2 million Tunisians receive a monthly income of under TD 4.17 (as published in the Development Plan).

TABLE 5.5

Monthly Salaries in Medenine
(in Tunisian Dinars)

Permanently employed in agriculture	$17,990 \times 20 =$	359,800
Others permanently employed (only men, without oil)	$24,373 \times 25 =$	609,325
Others permanently employed (women)	$1,315 \times 20 =$	26,300
Retired men	$990 \times 20 =$	19,800
Retired women	$310 \times 15 =$	4,650
Total		1,019,875

Sources: Personal calculations, and figures from the records of the governorate administration (1966 census extrapolated to 1973).

Social and Regional Behavior

The possible changes in the regional and social behavior of the oil workers are as important as the monetary flows. Without doubt, these changes are extremely difficult to record. It is even more difficult to retrace their origins in either the oil industry or other phenomena. A lack of time and certain organizational difficulties did not permit scientific selection of the persons interviewed on this subject. Therefore, the results are neither statistically correct nor incontestable as to their scientific content.[12] Nevertheless, they provide trend indications of important developments. Verification may also be found in parallel literature (though usually quite incomplete), or in the results of studies of other areas of the oil industry.

For many of the workers in El Borma who come from Southern Tunisia, the jobs in the oil industry are the first contact with the modern division of economy. This probably brings about an estrangement from traditional space and value concepts as well as the transformation of these men into an industrial worker in the Western sense.

These changes in behavior, which in the long run will also become spatially visible, must be registered. The most important characteristic is the employment of the considerably high incomes—at least for the Southern Tunisian standard. Of the 25 men interviewed (average age of 36), 19 came from the southern governorates of Medenine, Gabes, and Gafsa. All of them spent their incomes exclusively in their hometown and mainly on the construction of a house. Only five of the 19 natives did not yet have a house under construction or finished. One of them, however, had inherited a house and the four others belonged to the higher income group of TD 82 and more. Thus, the construction of a house is an important indicator of material prosperity. It appears to be the first step taken upon accumulation of some capital.

This is quite understandable considering the poor housing conditions in all rural regions of Tunisia. According to data from the 1966 census, 39.8 percent of all houses, in which an average of 5.2 persons lived (3.1 per room) were made of clay (Gourbis), another 2 percent were tents, and only 15 percent of all dwellings had running water.[13]

Another indicator of the changing mentality of the industrial workers and emigrants to Europe is evidenced by the fact that the new houses are generally not constructed in the traditional mode (haus), but rather as the European-style "villa," which, unlike the Arab model, faces the street.[14] A comparison of the various structural population indexes and their concomitant construction applications gives the picture shown in Table 5.6. No correlation between oil workers, the gainful employment rate, the emigration rate, and the

TABLE 5.6

Population Structure and Annual Construction Applications in Medenine

Delegation	Inhabi-tants (1972)	Gainfully Employed in Percent of Inhabitants	Emigrants in Per-cent of Inhabitants (1966)	Emigrants in Percent of Gain-fully Employed	Building Applications per 100 Inhabitants (1971-72)
Medenine	46,200	16.80	0.72	2.34	0.81
B. Kheddache	18,600	17.63	4.07	23.88	0.20
Ghomrassen	18,300	9.00	11.00	122.28	0.72
Remada	9,850	22.85	2.03	8.90	0.28
Tataouine	34,350	10.45	9.81	93.91	0.70
Ben Gardane	29,180	17.22	1.50	8.83	0.12
Zarzis	41,600	15.16	2.78	18.40	0.26
Djerba	79,750	16.13	4.34	26.70	0.20
Total	277,683	16.7	4.00	25.40	0.50
Tunisian total	5,200,000	26.35	3.46	13.14	—

Sources: Banque Centrale de Tunisie, Rapport annuel 1972 (Tunis, 1974), p. 4; and personal calculations from records of Service Travaux Publics–Bâtiments, Medenine Governorate.

house construction rate per 100 inhabitants may be drawn as a result
of the insignificant distribution of the regions of origin of the oil
workers. The emigration rate seems to be the only meaningful factor for the number of gainfully employed at home (with a negative correlation) and for the number of building permits per 100 inhabitants
(with a positive correlation). Of course, Medenine City, as a central
place with a considerable amount of functions in the administration and
supply sector, falls out of the correlation. On the other hand, the
low number of building permits should not mislead one to assume that
there is stagnation in the regions of Djerba, Zarzis, and Ben Gardane.
For several years now, continuous building activities have created
enough housing space to meet the increasing demand. However, in
the mountain regions such as Ghomrassen and Tataouine, due to the
recent influx of money from emigrating workers, opportunities to leave
the traditional underground "cave" dwellings and move into homes above
the ground are now being provided. This pent-up demand coincides
with better possibilities of emigration to Europe and, above all, to
Libya.

The oil industry, in spite of its few job offerings, figures among
the responsible factors for these changes, especially because its work
force brings about special and multiplying effects. The following
economic and social impacts on the region are thus visible:

The high income level allows the support of many people.
On the average, each of the interviewed persons with dependents (72
percent of total) supported 7.3 persons. Of all those interviewed
(unattached persons included), the average family contained 6.3 persons. These figures are above average. According to 1966 statistics,
each gainfully employed male in Medenine Governorate supported 5.9
people. If women and retired persons are brought into this calculation, the rate was 5.5 people per person with a regular income. In
all of Tunisia, however, only 3.8 were supported by one salary earner
on the average. This demonstrates that the high income allows a large
amount of dependents. It also shows in the ratio of the jobs provided
versus the paid salaries. Whereas the oil sector provided a mere
0.6 percent of the total of all sectors, the wages and connected enterprises contributed 1.5 percent to the total sum of salaries.

The high qualification demands of SITEP and its company-oper-
ated advanced training facilities mold the workers into responsible,
conscientious partners of a modern enterprise. This is accompanied
by a change in mentality closer to the modern industrial worker in the
Western sense. The alienation from the traditional rural society of
day laborers is evident: Only one of the 25 interviewed would not have
minded working again exclusively on his farm if he had no job in the
oil industry. The others would look for similar jobs, preferably in

the same region or in the rest of Tunisia or, with reluctance, in foreign countries (Libya and France). Thus, the oil workers, together with the workers in foreign countries, have pilot functions in their villages. Serving as models for their neighbors, they are making it easier for the latter to give up generally unprofitable agricultural work and to look for a better-paying industrial job.

The gradual integration into the modern profit-oriented economy is accompanied by an understanding of the necessity of a modern education and training. Most of the workers are illiterate, and want to, if they already have not, send their children (including girls), to modern schools. Not one of them was willing to let his son learn a traditional craft. The educational goal for their children is mediate or high school graduation. This aspiration for optimal education can materialize because of the higher income level. The families are able to move closer to the schools (new buildings are almost exclusively erected in new settlement nuclei including schools, shops, and so on). Furthermore, parents are able to maintain their families without child labor because of these relatively high incomes.

Effects on Regional Economy and Infrastructure

Aside from the already-mentioned effects on the labor market, no further effects on the regional economy can be depicted. The oil industry does not require any services from the regional enterprises except in the case of Sahara Comfort and some road construction and transportation firms. Occasional contracting of construction or painting jobs is rare since there is only one stone structure on the base (a community center with mosque, in existence since 1972). The personnel live in corrugated iron cabins. The entire food supply is either flown in or brought on special trucks from Tunis or Sfax. Food is not provided by the local agriculture, which, at present, is definitely not capable of supplying fair quality food with any regularity. (The same problem exists in the tourist center of Djerba.)

The oil industry also did not make any investments beneficial to the transport infrastructure of the region in general. The base is accessible only by air or special overland vehicles over a private dirt track, as the paved road ends at Remada. The recently paved private airport runway of El Borma is equipped to handle Caravelles and is landed upon twice a week by a chartered Tunis Air plane. The company, however, participates in the maintenance of the routes GP 19 and MC 101 between Kambout (south of Remada) and Djeneien, which are, in contrast to indications on official maps, only gravel roads. The El Borma base is a private settlement not open to the public. A permanent base is neither planned nor to be planned, as it is already lo-

cated in territory hostile to permanent settlement. The lack of usable water (even the deep ground water is brackish) is not the only
important reason to abandon thoughts of a permanent settlement (as
planned in Algeria). Except for the oil company, the contractor's
bases, and a military base (20–50 soldiers—founded in 1961 before
the arrival of the oil industry) to guard the Tunisian-Algerian border,
there are no further settlements in this area. The base will be completely deserted in 10–15 years when the field will be exhausted. The
first indications of exhaustion are already evident.

DOULEB

The Saharan oil deposits raise very specific problems. The
high expenses in exploration, the unfavorable geographic conditions
for settlement, and the lack of all other factors needed for regional
development of any kind are the reasons for marginal, quickly erected,
and soon-deserted installations. The question whether this marginality
and the usually lacking linkage to the regional economy depend on the
characteristics of the natural setting or those of the oil industry leads
to a comparison of facilities in different landscapes. Two other Tunisian oil fields located outside the Sahara will serve this purpose.
The oil field of Douleb, located in the woods of Djebel Semmama
(at an elevation of 1,314 meters) 38 kilometers northwest of Sbeitla,
is being exploited by the French-Tunisian company SEREPT. It aver-
ages an annual output of 200,000 tons, in addition to 20,000 tons from
nearby Tamesmida and Semmama. Production, in 1974, totaled
255,000 tons. The initial discovery of the fields was in 1966 and production started in 1968. Without additional pumping stations, the
crude oil, including the Tamesmida/Semmama output, flows to Skhira
on its own through a pipeline (175 kilometers in length and 6 inches in
diameter). The problems in the economy of Kasserine Governorate
are comparable to those of Medenine in spite of the varying natural
terrains. The Djebel Semmama is covered by Mediterranean-type
woods, a snow cover lasting several days being not rare in the winter;
whereas Sbeitla and Kasserine are already located at the northern
border of the Halfa grass steppe of the central highland, which more
and more is being cultivated with orchards and grain.

Labor Market in Kasserine Governorate

Of an estimated population of 258,000 persons (January 1973),
of whom 88 percent live in rural areas, only 54,300 men (42 percent
of all men, 21 percent of the population) are economically active.[15]

TABLE 5.7

Nonagricultural Jobs in Kasserine, 1973

Industries (including small and medium-sized firms)	3,012
Building trade	1,750
Crafts	565
Trade	2,392
Administration	2,977
Total	10,696

Source: La situation de l'emploi dans le Gouvernorat de Kasserine (Tunis: Ministry of Planning, 1973).

Of all of the economically active men (42,700 or 78.6 percent in agriculture; 11,600 or 21.4 percent in the nonagricultural sector), 28,000 or 51.5 percent were either unemployed or underemployed. Here, too, the emigration, which officially and unofficially reduces the work force by approximately 500 persons each year, is a significant factor. Modern industrial jobs are almost exclusively provided by the national enterprises Société Nationale Tunisienne de Cellulose and Société Tunisienne de Papier Alfa, with a total work force of 930 workers. Additional jobs are provided by other mining businesses, so the modern industrial sector employs 1,432 people (65 women). Of these, only 39 persons are directly employed by the oil mining industry. Altogether, the nonagricultural jobs in Kasserine Governorate are distributed as shown in Table 5.7. Predominantly traditional, nonproductive sectors still exist.

Monetary Flows

SEREPT employs 37 persons (including Temesmida/Semmama oil fields and watchmen of the pipeline); another four are employed by contracting enterprises for catering facilities; and six truck drivers are employed by the national transport company (STM) for driving oil trucks from Tamesmida to the pipeline. Irregular ancilliary jobs (repairs, upkeep on the buildings) require another approximate 2,000 workdays, which corresponds to a permanent employment of eight workers. So, according to the number of employees, the share of jobs in the oil industry among the total of industrial jobs (1.8 percent) and the nonagricultural jobs (0.5 percent) is minimal.

The monetary flows caused by the distributed income are also only of trifling importance. The incomes earned in the oil industry are relatively high due to the higher qualification requirements of the work force and the often-unfavorable working hours. The Halfa factories SNTC/STPA pay TD 16.75 per month on the average and 70 percent of the work force of the crafts and small industry have a monthly income of under TD 15 per month, while the salaries paid by SEREPT reach TD 50 to 65; for unskilled laborers and watchmen TD 30 to 35. The origin of the workers shows a regional differentiation according to their qualifications. Only one of the 12 foremen and intermediate foremen comes from Kasserine Governorate; two live in Sbeitla. Of the 20 registered workers and unskilled laborers, only three come from outside the region. Eighteen live in Sbeitla as well as in little villages near Douleb. Adding the contractors, who are not registered here, TD 1,925 per month are paid to local people.[16]

In comparison with the estimated TD 51,747 paid by the other branches of modern industries, the oil industry accounts for only 3.5 percent. However, it also means that the oil industry figures in only 0.2 percent of the total sum of wages that are available in that region including agriculture (excluding money transfers from emigrants working abroad). So, even the financial significance of the oil sector is minimal.

In addition to the salaries are reimbursements for agricultural property used now by the Douleb-Skhira pipeline. These were generally small lump-sum payments and they are certainly insignificant in contrast to other oil-producing regions. Moreover, the management was not able to give exact figures of these payments.

Regarding the social and economic behavior of the oil workers, the results attained in 16 interviews (total of the base's day population) are similar to those of El Borma. In spite of much personal mobility in looking for a job, the ties to native villages are a determining factor. All of the interviewed live in their native villages with their families (except for two who migrated from Gafsa and Zarzis to Tunis). Practically the entire income is spent there, predominantly for the construction of a house. Eleven of the 16 interviewed (average age 31) had erected a house or had one under construction, and only one of these was a traditional farmstead (hansir). Although jobs are accepted far away from the hometown, especially by persons in the higher income brackets, the family remains in the native village where the workers eventually want to retire. At most, moves within the region (from rural areas into the next small town) are considered.

The multiplicatory effects described in El Borma are noticeable as well:

A worker supports an above-average number of family members (7.2 persons). The absolute numbers vary from 1 to 15.

The change in mentality is even more important. Eight of the 16 interviewed came from farming families. The parents of five other workers also had traditional occupations (fishermen, charcoal burners, peddlers). In spite of this, only one of them would perhaps be willing to return to agriculture if the job in Douleb were no longer available. All of the others declared that if this should happen, they would look for a job in the modern sector, considering migration within Tunisia or even emigration to Libya. Although primarily illiterate, they care a great deal about a school education for their children (including girls). Here, too, an attachment to the modern labor-division economy occurs.

No effects on the regional economy could be found aside from the paid sum of wages. There were also no impacts on the agriculture. The work force of the oil industry forms a market too small to be considered. Aside from the few local craftsmen employed as contractors, influences on local trade and industry are also not existent. The approximate 15,000 cubic meters per day of Dolomite gas (one-third of which is nitrogen) brought to light in the course of the oil extracting processes are completely burned. The once-considered use of natural gas for the Halfa processing factories does not seem realistic at all because of the poor heating power of the gas and the already foreseen exhaustion of the natural gas deposits.

The drop in production will probably begin in 1978. Further production will be stopped when the expenses exceed the profits. The regional infrastructure has not been improved in any way. The Douleb base can be reached on a route partly paved with gravel and at times is accessible only by vehicles with four-wheel drive.

Considering all these factors, the oil industry remains only marginal for that region. Once production stops, there will be no traces left behind except for the solid buildings at the Douleb base. With regard to the fact that the annually produced oil in the Douleb/ Tamesmida base has a foreign trade value of almost TD 10 million, it can be realized that the strong financial effects of the oil production are not accompanied by effects on the region's economy.

SIDI EL ITAYEM

Sidi El Itayem may serve as the last example for an oil field in the temperate climate zone. Discovered by the French CFTP, the field started production in 1972. Since then Tunisia has been an equal partner in the venture. Located in the middle of olive groves, four kilometers off the road GP 13 from Sfax in the direction of Menzel Chaker, the site is visible only because of the gas torch. Original production plans were to be 400,000 to 500,000 tons per year, but as of 1975 had not exceeded 200,000 tons.

The administration of the field in the town of Sfax employs five persons. Another 17 work in shifts as operators. Of the total of 22 workers, four are French, the others come from Sfax Governorate. Wages totaling approximately TD 720 per month or TD 8,640 per year are being paid in the region, plus an additional TD 2,000 to 3,000 for contractors. The oil products are transported via pipeline to Skhira. There are thoughts of using the produced gas for the generation of electricity in Sfax, as each ton of oil produced is accompanied by 50–60 cubic meters of gas. The production would be high enough to operate two generators of 5 megawatts each. The national gas and electric corporation STEG has recently entertained the idea of a project of building a gas turbine electricity generation plant at Sfax, powered by gas from Sidi El Itayem. However, no figures as to its capacity were disclosed.[17]

No effects on either the labor market or the regional economy and infrastructure can be stated. The management cited that the price level rose with the discovery of the field because each of the merchants hoped for additional profits due to an oil boom and the arrival of wealthy foreign engineers. This, however, should not be taken seriously. Until now, there has been neither an addition to the energy supply facilities nor other improvements of the locational qualities of the city of Sfax. With approximately 250,000 inhabitants (243,810 estimated in 1971), it is the second largest city in the country. Here, too, the oil industry remains isolated from the other industries.[18]

ASHTART OFFSHORE FIELD

This isolation can be seen more distinctly, not only by the appearance, but also because of the characteristics of offshore drillings. The logistics base of Ashtart is Sfax. Supply and work force are generally provided by foreign companies; only the unskilled laborers come from the local region. This effect is further added to by the fact that the permits for offshore drillings allow foreign companies to spend up to 85 percent (otherwise 70 percent) of the investments outside of the country. The Ashtart field was discovered in October 1971 by the French company Aquitaine Tunisie in an equal-partner association with ERAP. The field is situated 80 kilometers southeast of Sfax at a depth of 66 meters. Originally, production was estimated to reach 1.2 million tons per year but, in fact, it reached 1.5 million tons in 1974 and may eventually reach 2.5 million tons. Exact figures of the reserves are not yet indicated, but they seem to be between 15 and 30 million tons.

Drilling and production platforms rise 85 meters from the sea bottom; 5,310 tons of steel were used for their construction. The produced oil is stored in a 70,000-liter floating tank and then pumped to tankers. The gas is burnt or used for the power plant covering the internal demand.

Altogether, approximately 50 persons are working on the "island" itself. Another 50 work in the administration and supply departments in Sfax. Also, there are two ships and one helicopter for communication between Ashtart and Sfax. Unfortunately, no further data concerning the structure of the work force and the possible amount of regional impacts could be obtained. It can be gleaned only from the literature and publications.[19]

NOTES

1. Alger Réalités 1, no. 5 (1972): 56; and "Potentialités et perspectives de l'exploitation pétrolière," L'Action (Tunis), March 7, 1975.

2. Industries et Travaux d'Outremer (Paris), April 1976, p. 286.

3. Information from Délégué Social du Gouvernorat de Medenine.

4. H. Attia and M. Rouissi, "Structures agraires et développement dans le Sud tunisien," Annales algériennes de Géographie, Special number (Colloque de Ouargla), September 25-26, 1971, pp. 15-34 (Algiers, 1972). For the Djerba Island tourism, see G. J. Tempelman, "Tourism in South Tunisia: Developments and Problems in the Djerba-Zarsis Region," Tijdschrift v. Econ. en Sociale Geografie (Amsterdam) 66 (1975): 21-34.

5. Les villes en Tunisie (Tunis: Direction de l'Aménagement du Territoire, 1971), pp. 457-60.

6. Remark of the Délégué Social du Gouvernorat de Medenine, May 1973.

7. Random spot check of 25 SITEP workers, May 1973.

8. See also Chapter 4.

9. According to L'Action, May 17, 1974.

10. M. Seklani, "Le champ d'intervention du salaire minimum s'élargit à l'agriculture," Conjoncture (Tunis), July-August 1974, pp. 24-26.

11. See IVe Plan de développement 1973-1976 (Tunis, 1973), p. 180.

12. This is even more true for our studies in Algeria (see Chapters 6 and 7).

13. "La situation de l'habitat," L'Action, September 26, 1973.

14. S. Pompei, "Problèmes d'urbanisme dans le Sahel," Cahiers de Tunisie, nos. 47–48 (1964), pp. 147–63.

15. These and the following figures from La situation de l'emploi dans le Gouvernorat de Kasserine (Tunis: Ministry of Planning, 1973).

16. Calculated from SEREPT data.

17. "Le programme de developpement de la STEG," Marchés tropicaux, December 26, 1975, p. 3676.

18. For further regional data from the Sfax Governorate, see Chapter 8.

19. "Le gisement d'Ashtart," ELF Bulletin mensuel d'Information (Paris), December 1973, pp. 3–11; "Apercu sur les résultats du groupe ELF–Aquitaine en 1973," ELF Bulletin mensuel d'Information, January 1974; "Mise en production d'Ashtart," Petrole Informations (Paris), December 13, 1973, pp. 15–16; "Un baril est un baril," Le Monde, December 9–10, 1973.

CHAPTER

6

EFFECTS OF OIL
PRODUCTION IN
ALGERIA

The conditions of the hydrocarbon-producing fields in Algeria are different from those in Tunisia. Historically, this area of the Sahara, delimited by the Wilaya Oases (in its border until July 1974), had been neither a marginal region nor was it uninhabited. It had important transit functions within the manifold trans-Sahara trade connections until the colonization period, when trade gradually became exclusively Europe-oriented. The existing or easily accessible ground water supplies permitted settlements that had assumed important supply duties during this trade period. A good network of settlements developed in this way in the Northern Sahara region, limited by the mountain range of the Sahara Atlas in the north and Ouargla, maybe El Golea, in the south. Road distances between the urban settlements never exceed 170 kilometers (except for Ghardaia, which is 240 kilometers from El Golea). The hydrocarbon fields of Hassi R'Mel (in the vicinity of Ouargla) and Hassi Messaoud with their technical and residential facilities are fitted into this network. The In Amenas oil base, located 800 kilometers by road from the nearest city (Ouargla) is already in the "Great South," which is not exactly unpopulated but offers only a few natural and economic assets for permanent settlements. From its physical setting the Northern Sahara is certainly a potential producer region even for agricultural products. However, the great distances to the consumers make a profitable marketing of the products in the framework of a market economy nearly impossible.

What changes did the arrival of the oil industry cause in the economic, social, and regional structure of this area? Did this industrial branch, which radically changed the financial structure of the country, affect the region itself? A study of several hydrocarbon fields, which are seen within the total complex of the region, attempts to answer these questions.

89

FACILITIES AND PROBLEMS IN
THE WILAYA OASES

As already stated in the introduction to Algeria (Chapter 3), the population problems are similar to those in Tunisia: high growth rates and high rates of under- or unemployment. A few favored locations on the coast with comparatively balanced economic growth are contrasted by rural, marginal regions with pressing population demands. Unfortunately, the Algerian statistics make it impossible to record the internal migrations of the population. Also, the current population growth is only extrapolated for the entire country and thus figures vary considerably. Whereas the 1974 Statistical Yearbook registered 13,523,144 residents in 1971, 13,954,915 in 1972, and 14,387,070 in 1973,[1] a different official publication reports as many as 14,643,700 persons in April 1971.[2] The difference could possibly be the addition of the emigrant workers in foreign countries—274,668 persons, according to a 1966 census, and an estimated number of 750,000 at present—plus the foreigners living in Algeria (1966: 188,194 persons). Table 6.1 is a comparison, disregarding these differences, of the 1966 and 1971 populations of the individual Wilayats (the growth rates only serving an internal comparison).

According to the figures in Table 6.1, the Wilaya Oases, together with the Wilaya Saoura, that is the entire Algerian Sahara, have the lowest growth rates in Algeria. The reason for this negative development is certainly not so much the situation of the labor market as it is the climatic impediments and the distinct marginal location (in comparison to North Algeria). Remarkably, in 1966, 23.7 percent of the Oases population (Algerian average: 19.8 percent) were economically active, of which 21.8 percent were from the producing sector (Algerian average: 17.9 percent) and 26.7 percent in the service sector (Algerian average: 24.5 percent).[3]

The comparatively favorable ratio between work force and jobs improved even further—at least, in the statistically recorded "modern" sector: The number of jobs in this sector had increased by 27.1 percent in Algeria as a whole between 1969 and 1971, but the Wilaya Oases showed a growth rate of 39.2 percent. This means that there were 3.1 jobs in the modern sector per 100 inhabitants (Algerian average: 2.3). Since 1966, 15,000 new jobs, of which 11,000 were permanent, have been created so that the unemployment rate could be lowered from 30 to 12 percent. The overall investments within the special program (see Chapter 3) totaled DA 410,066,830 since November 29, 1966.[4]

However, evaluation of the oil industry's impact on monetary flows and the labor market has to be based on assumptions. In 1971, of the 17,826 industrial workers (3 percent of the total population),

TABLE 6.1

Rate of Population Growth in Algeria on Wilaya Basis, 1966–71

Wilaya	Growth in Percent, 1966–71	1971 Population in 1,000s	1971 Density of Population per Square Kilometer
Algiers	26.2	2,079.0	665
Annaba	16.9	1,108.8	45
Aures	16.7	893.8	24
Constantine	16.9	1,772.2	89
El Asnam	26.0	992.9	77
Medea	24.9	1,087.7	22
Mostaganem	22.6	953.6	84
Oases	13.5	572.0	0.5
Oran	22.9	1,174.9	71
Saida	22.8	290.6	5
Saoura	13.3	238.8	0.3
Setif	17.2	1,447.0	79
Tiaret	22.1	442.3	16
Tizi–Ouzou	25.7	1,045.1	172
Tlemcen	22.7	545.0	67
Algerian total	21.0	14,643.7	6

Sources: Annuaire statistique de l'Algérie 1972 (Algiers, 1973), p. 17; and Algeria in Numbers 1962–72 (Algiers: MIC, 1972), p. 12. No population data on Wilaya basis have been published since.

8,437 (52.5 percent of the industrial workers) were employed in the oil industry (prospect and production). This number seems to be small. In our investigations we have to take into consideration an additional group of one-third to one-half of the workers in branches closely connected with the oil industry. A good share of the construction and transport businesses ("contractors") belong in this category. The Wilaya Oases have an above average number of these groups totaling 8,276 persons. Altogether, only 1,129 persons (6.3 percent of the industrially employed) are employed in other industrial branches.

Table 6.2 presents our calculations (with all cautions of such calculations based on the statistics of a developing country) of the sums of money annually at the disposal in the Wilaya Oases (population

TABLE 6.2

Estimates of the Available Incomes of Active Population in the
Wilaya Oases, 1971–73
(one year)

Sector	Number of Employed	Monthly Wage in DA	Total Amounts in DA	Percent of Total
Agriculture, socialist sector	900	—	3,000,000	0.4
Agriculture, private sector	69,500	165	137,120,000	19.3
Building trade (not registered)	8,000	400	38,400,000	5.4
Trade, services, and other activities	33,500	500	200,100,000	28.1
Building trade and transport apart from oil industry	4,138	600	29,793,600	4.2
Building trade and transport in oil industry	4,138	600	29,793,600	4.2
Oil industry	8,597	1,300	134,113,200	18.9
Oil exploration	2,490	800	15,072,000	2.1
Administration	11,440	900	123,552,000	17.4
Total amount	142,703		710,944,400	100
Oil industry total	15,225		178,978,800	25.2

Sources: Personal calculation based on official data, partly
extrapolated. For the oil-exploration sector, see Chapter 4. Partly
differing data of average wages are found in Annuaire statistique de
l'Algérie 1974 (Algiers, 1975), p. 212.

figures of 1971, income level of 1973). Upon closer inspection these
numbers seem to be realistic for they result in a 24.5 percent activity
rate of the population and a per capita income of DA 1,216.5 per year
(approximately DA 100 per month). Even though the work force of
the construction business and the administration, as well as the earned
wages, are possibly estimated too high, an even stronger domination
of the oil sector can be assumed.

More recent figures, which could not be included in Table 6.2
because of the lack of data for population development and agricultural

employment, prove that these indications are correct. As most of
the personnel of contracting enterprises are now included in
SONATRACH, employment in the oil sector had risen from 11,227
persons in 1972 to 16,084 in 1973 and 20,149 in 1974. Due to more
appropriate statistical data, employment in the building sector also
grew from 8,674 persons in 1973 to 9,317 in 1974.[5]

Of the available incomes, 25.2 percent are earned directly in
the oil industry, although only 10.6 percent of the overall work force
are employed in this sector. This is caused by the enormous differences in the income levels. A farmer earns approximately DA 150
to 200 per month; a helper in the second sector earns DA 400 to 600;
an untrained oil worker (the usual status of a worker from the South),
however, earns DA 1,200 to 1,500 per month.

An overall examination of this kind cannot be complete because
of the size of the region and the multitude of factors affecting and
caused by the oil industry. The following chapters provide—aside
from information on the individual oil fields and bases—an overall
view of such factors and related problems.

HASSI R'MEL GAS FIELD

Facilities

Two thousand billion cubic meters of the Algerian total natural
gas reserves of 3,570 billion cubic meters (1975, representing 5.6
percent of the known world reserves) are found in Hassi R'Mel, * the
largest natural gas deposit in service in Algeria.[6] The field was discovered in 1956 by the French companies S.N. REPAL and CFPA.
Production started in 1961 (231 million cubic meters). Since the gas
liquefaction plant in Arzew began operation in 1964, production has
increased to 2.6 billion cubic meters (1968), to 2.9 billion (1971),
and to 5.6 billion (1973). By January 1, 1974, a total of 33 billion
cubic meters of natural gas and 5.6 million tons of crude oil (which
equals 210 grams of condensate per cubic meter of natural gas) had
been produced. So far, the production has been relatively small,
resulting from rather limited market outlets. The two gas pipelines
to Arzew (1961, 28-inch) and to Skikda (1972, 40-inch) were supplemented in 1974 by a 40-inch pipeline. The oil is transported to Hassi
Messaoud through an 8-inch pipeline. The producing capacity, which
in mid-1973 was already 15.7 billion cubic meters, is projected to

---

*In Arabic, Hassi(er) raml, which means sand well.

reach 43 billion cubic meters per year in 1980. An additional planned
gas pipeline will extend to Oued Isser (east of Algiers), which is the
projected location of the country's third natural gas liquefaction plant.
Gas treatment at Hassi R'Mel will be intensified by a new unit to be
completed in 1967, which will produce 18 billion cubic meters of dry
gas and 3.5 million tons of condensate.[7] The gas has a strong content
of methane (84 percent) and a high caloric value of 9,400 kilocalories
per cubic meter. The life expectancy of the deposit is estimated to
be 50 years.*

Labor Force

SONATRACH, which has owned and exploited 100 percent of the
deposit since 1971, offers 140 working positions that equates to 285
jobs because of the operation in shifts.

The employed are associated with the following services: head
department (medical care, occupational training, external contacts)
12; administration and supply department (including catering) 59, of
whom 37 are in supplies and 5 in the amusement department (Loisirs);
security (firemen, watchmen) 20; general services (transport, repairs, storage, paint) 73; gas production and geology 39; gas processing plant (current operations) 48; technical service of the gas processing plant 29; additional employees (highly paid engineers outside the
organization chart) 5. Of these workers, 65 percent come from the
Algerian north, the rest from the south. Only three persons are foreigners. So far, all of the employees live in prefabricated wooden
houses and spend their pastime (replacements in 6/2 weeks rotations)
in their native towns.

Because of the life expectancy of the field, the company has
started a building program in which 120 solid residential houses for
four- to six-member families are to be built (by mid-1973, 40 had
already been finished). The 12 families presently living in prefab
bungalows will be the first to move in. That a program is planned
for the construction of accommodations for 700 unattached workers
indicates a future increase in jobs. This corresponds with the projected increase of production as well as to a concentration at
SONATRACH in those services thus far partly provided by contractors.

---

*For general questions concerning the Algerian gas industry,
see Chapters 12 and 13.

Ancillary Services and Supply

Five persons are employed at the energy center of the state SONELGAZ (gas and electric company). This center, with a capacity of 10 megawatts, supplies the settlement and the industrial facilities as well as Ghardaia (until 1973). An expansion of the facility is planned.

The public administration of Hassi R'Mel is only rudimentary as it is not a public settlement. Employed are three teachers (for the 33 children of the residential families), 12 policemen, and one head administrator (appointed by the Wilaya's administration) with five employees. A health center with three male nurses is maintained by SONATRACH. In addition, ancillary services are provided by the contracting construction enterprises, which, however, maintain their own bases. The exact number of employees could not be obtained (see Chapter 4 for further information on the contracting drilling company ALFOR).

The only facility open to the public is the hotel of the French Compagnie Hôtelière Saharienne (CHS), which, with an average personnel of 20 persons, provides 84 beds in mobile living cabins. In 1972 the hotel registered 6,000 overnight stays. The workers of those contractors who do not have a base in that region usually find their accommodations here. This adds another 22 workers to the yearly average. Tourism is practically unnoticeable, especially since the base is off the transit routes and the facilities are not public. Taking into account the 70 workers employed by the construction enterprises, the work force in the entire Hassi R'Mel region associated with natural gas totals approximately 425 persons.

All the food for SONATRACH bases comes from Algiers, except for occasional purchases of meat from the passing nomads (amounting to DA 1,000 per month). In contrast, the CHS purchases on the average 80 percent of its food supply (value of DA 70,000 per month) in Ghardaia, whereas formerly all goods came from Algiers.

Associated Processes

Hassi R'Mel is not now and will probably never become a public settlement. The location is marginal in comparison to the considerably more attractive, traditional centers of Berriane (85 kilometers away), Ghardaia (130 kilometers), and Laghouat (120 kilometers). This is the reason that there are no plans for an expansion of the settlement or the settling of persons not connected with natural gas (such as nomads). All in all, the base is completely isolated, provides its own energy, and does not "import" anything from the surrounding region—not even an appropriate share of its work force. So far, the only infrastructural effect was the completion of a 25-kilometer asphalt road off of the Algiers-Ghardaia route. However, in the future, Hassi R'Mel will also produce electrical power for the region. In 1975, installation work began on four turboalternators with a capacity of 20-25 megawatts each. They will furnish electrical energy to the Laghouat and Ghardaia regions.[8]

HASSI MESSAOUD OIL FIELD

In contrast to the other petroleum fields and bases, Hassi Messaoud* aroused the interests of scientists at an early date. From its discovery in February 1956 by two French companies (S.N. REPAL and CFPA) and until its nationalization on February 24, 1971, S.N. REPAL was the operator in the northern part and CEPA in the southern part. With an estimated reserve of 3 billion tons, of which 500 million tons of oil can be extracted, and with an annual output of 20 million tons (12 in the north, 8 in the south), the life expectancy of the field is believed to be almost 30 years.

Facilities

The splitting of the field by the concession boundaries resulted in two settlements: the base "February 24" (Base du 24 fevrier, formerly Maison Verte—CFPA) in the north, and the Base Irara in the south. Each of the two bases has its own processing plant.

Hassi Messaoud and the large pumping station in Haoudh el Hamra are becoming more and more the focal point of the developing Algerian pipeline network. Pipelines lead to Bejaia (built in 1959, 24 inches in diameter, original capacity 4.6 million tons, presently 17.5 million tons per year); Arzew (1966, 28-inch, 805 kilometers, original capacity 14 million tons, presently 22 million tons); Skikda (from Mesdar, 1973, 34-inch, 637 kilometers, capacity at the first stage 22 million tons, at the final stage 30 million tons); and from Ohanet (In Amenas—1961, 30-inch, 518 kilometers, capacity 8 million tons). Two pipelines from several local oil fields also lead to Bejaia: El Agreb (8- to 10-inch, 105 kilometers), and from Rhourde el Baguel (14-inch, 109 kilometers). In addition, the condensate/oil gas (liquefied petroleum gas, LPG) pipeline to Arzew, which started operation in the middle of 1973, has 300 kilometers of a 12-inch pipe

---

*In Arabic, Hassi Mas^{c-}ud, which means happy well.

to Hassi R'Mel and thereafter 505 kilometers of 16-inch pipe. Its initial capacity was approximately 0.85 million tons of oil gas and 1 million tons of condensate per year.[9]

Between the two camps and Haoudh el Hamra are grouped the different technical installations, notably the separation, stabilization, compression, and LPG treatment units. The very complex physiognomy of the base shall not be discussed further as it has already been described in a study (from early 1971) and also appeared cartographically in several essays.[10]

Labor Force

As a result of its size and central location—in comparison to the other Algerian deposits—Hawwi Messaoud was developed early as a residential base as well. Its functions, which exceed those of a pure extraction base, made it the biggest oil city in Algeria. The present population is between 5,000 and 6,000 after its peak of 8,500 during the developing period. In early 1971 the oil companies S.N. REPAL and CFPA directly employed 675 and 856 persons, respectively. The two pumping stations of Haoudh el Hamra employed 150 men.[11] The rest of the inhabitants consisted of family members, spontaneously settling nomads, and workers employed by the contracting enterprises. After nationalization, 1,721 persons were employed by SONATRACH (August 1, 1971) of whom 60 were foreigners with consulting functions. Now, approximately 2,500 persons are employed by SONATRACH, as nationalization brought about an increased integration of the contracting enterprises into SONATRACH. The origin of the employees of the oil companies themselves was as shown in Table 6.3 (using as an example CFPA, in the beginning of 1971).

A spot check of all of the Service Puits personnel in May 1973, after almost all the foreigners had left, produced the figures in Table 6.4. Apparently, the number of Southern Algerians had not changed, whereas the vacancies caused by the departure of the French were filled mainly by Northern Algerians. The clear underrepresentation of Southern Algerians in higher income brackets parallels the observations made in other oil fields, where employees from the local area face the same difficulties as a result of less education and training. The exclusively dominating power of, in this case, Algiers,[12] affects traffic connections, the food supply, and the like. Since the entire administration and training facilities are concentrated in Algiers (the Institute Algérien du Pétrole branch in Hassi Messaoud recruits its probationers predominantly from the North), the Sahara inhabitants are constantly—often unintentionally—at a disadvantage.

TABLE 6.3

Origin and Qualification of the Employed in Hassi Messaoud, 1971

Qualification	Foreigners (French)		North Algerian		South Algerian		
	Abso-lute	Per-cent	Abso-lute	Per-cent	Abso-lute	Per-cent	Total
Foremen and executives	109	12	18	2	3	0.3	130
Specialists	82	9	52	6	49	6	183
Workers and helpers	12	1	79	9	471	54	562
Total	203	22	149	17	523	60	875

Source: G. Corna Pellegrini, Per una geografia delle città pioniere—Hassi Messaoud, pub. no. 4 (Milan: Catholic University of Santo Cuore, 1974), app., map 33.

Ancillary Services

Correct figures of the number and structure of the contracting firms could not be obtained from publications or other sources in Hassi Messaoud. Without doubt, the number of employees of these firms is subject to extreme fluctuation depending on the demand for their services, the political situation, and so forth. Particularly after the nationalization, some firms stopped working and their personnel were taken over by SONATRACH (all catering enterprises and facilities for the oil workers). The public and semipublic firms (trucking, drilling) taking their places have their seats and recruiting offices in Algiers; they send their personnel to the south only for short periods of time and on a rotating basis.

So, today the industrial zone is in part a gigantic junk yard with only a few indications of former activities. Found here are a bottled gas depot, a drugstore (wholesale), a depot of the road department, a post office, a hotel, SONELGAZ (National Gas and Electric Company), Air Algérie, Schlumberger (contracting firm, deserted), Flopetrole (contracting firm), a butane gas depot, two gas stations, SICLI fire extinguishers, auto repair shops, ALFOR/ALDIA (drilling firm, SONATRACH branch), SNTR (public trucking company), electric appliances enterprise, building contracting firm (private), a carpentry (deserted), a welding firm (deserted), and SONADE (water supply).

TABLE 6.4

Origin and Qualification of the Employed in Hassi Messaoud, 1973

| Qualification | North Algerian | | South Algerian | | |
	Abso-lute	Per-cent	Abso-lute	Per-cent	Total
Executives	9	5	1	0.5	10
Specialists and foremen	18	10	10	5.0	28
Workers and helpers	53	29	93	50.0	146
Total	80	44	104	55.5	184

Source: Personal analysis of SONATRACH records, 1973.

The French contracting firms in particular, which had been operating in this area before the nationalization, left almost all their equipment behind—the transport back to France was not worth the expense. A mere attempt to estimate the number of employees in the industrial zone is impossible.

The fate of the private transport enterprise RBTOK (Rouigi Bachir et Oulad Kouider) clearly depicts the company's intertwining with local enterprises and the rising oil industry and then its subsequent decline.[13] This is also indicative of the course of the money flow generated by the oil industry. Two merchants from Ghardaia founded RBTOK in 1948 as a Sahara trucking company and in 1957 set up operations in Hassi Messaoud. The present facilities were built in 1962 (garages and lodgings), and at the peak of its expansion the Hassi Messaoud branch employed 156 workers, with 56 heavy-duty trucks and 25 lighter vehicles. Upon the establishment of a state transport monopoly for long-distance trucking, trucking firms were granted a reprieve and were allowed to operate until their vehicles wore out. The purchase of new vehicles was—officially—prohibited. In 1973 RBTOK had seven trucks, three mobile cranes, and three buses, almost all of them more than ten years old.

Regarding the origin of RBTOK's employees (May 1973), it is seen (in opposition to corresponding statistics on SONATRACH) that all employees in the Hassi Messaoud branch came from the Sahara. They are predominantly from the Mzab (29 persons), where the owners lived; and also from the Great South (five from Timimoun, two from El Golea), although the 20 drivers and 12 mechanics belong to the category of specialists. Two families live permanently in Hassi Messaoud. As there is no future in the transport business, the owners built a radiator factory in the new industrial zone of Ghardaia (see Chapter 7) in 1972. Eventually the facilities in Hassi Messaoud will be changed into a radiator repair garage. In any case it seems that there is no future in Algeria for private enterprises and especially for those acting as contractors for the oil industry.

Settlement and Supply

Hassi Messaoud was the only oil settlement of those visited with a permanent settlement of families being directed by the oil companies and consisting of spontaneously settling nomads.

The administration center, together with the adjoining residential area, forms the nucleus of a genuine, public settlement with free access to everyone. Found here are an administration building (Hassi Messaoud is administered as an industrial unit within the community of Ouargla by an "administrateur délégué"; the formation of a politically independent community has been planned for years), an elementary school, and a health center with a medical doctor. Hassi Messaoud has the following facilities of the tertiary sector: police headquarters, post office, pharmacy, travel agency, and three retail stores for food, household equipment, electrical appliances, crafts, gifts, clothing, shoes, stationery, and books.

This modern shopping center, and the few traditional Arab-style shops of the Cité Résidentielle, are attached to the former "European" living quarters. In 1971 this section contained 50 villas; in the meantime another 170 have been added.[14] In contrast to the usual living cabins of the two residential zones, the families can live here comfortably all year round. If possible, however, the mid-summer season is spent in the North. The solid state, and fully air-conditioned villas (with 70 square meters of interior space for average families) are provided by the company for its employees in higher income brackets without charges (including utilities and care of the gardens). This results in a stronger permanency of the employees and also saves on compensation payments for the separation from and monthly flight back to their families. Several of the resident families do feel comfortable and conceive living there until their retirement (three out of four interviewed who live in Hassi Messaoud with their families). In contrast, members of the lower income groups are generally not willing to settle with their families in Hassi Messaoud. They do not want to break ties with their hometowns and especially fear the contact of their wives, generally raised according to the old traditions, with the modern industrial world. Only two of nine interviewed persons with income levels of DA 1,500 or below were willing to live with their families in Hassi Messaoud, in contrast to four of five inter-

viewed with a higher income level. However, none of the interviewed
could imagine spending their retirement there.

This intensified settling of specialists in Hassi Messaoud and
the expansion of Cité résidentielle to a real urban city are obviously
promoted by the Algerian government. In August 1974 an international
tender of SONATRACH was published for construction in 1975 of 1,080
apartments with an interior space of 150,000 cubic meters. These
are to consist of prefab units and are to be constructed by a prefabri-
cating plant in Hassi Messaoud. Also planned was an enlargement in
massive construction of the social, cultural, sporting, and shopping
facilities encompassing 65,000 cubic meters of interior space. Although construction had not started at the time of writing, the project
has been further enlarged. In 1976 it was planned to provide an additional living space in collective and individual buildings for 2,000
families, to be completed by 1980. They will be supplied by new
shopping centers, sporting grounds, and an open-air theater. It is
even reckoned that, with all children being in school by 1980, more
than 2,100 places in primary schools will be needed. A technical
high school is projected, too.[15]

Nomads have settled spontaneously since the foundation of the
oil settlement in Hassi Messaoud, located at the old caravan route
from Touggourt to Zaouia el Kahla and the Hoggar mountains. This
settling process started during the construction of the base. In 1959
the number of squatter settlers was estimated at 800; they were living
in derelict prefab cabins or in huts made from junk (tin, wood, card-
board) similar to the well known "Bidonvilles."[16] In 1968 their number was estimated to be 300 to 400 families. The number of 30 families given to us in 1973 was certainly too low.

All of these nomads, who preferred to settle at the outlets of
the sewage pipes of the Cité résidentielle, take advantage of the various, if often poor, earning opportunities offered by the industrial
settlement. For one thing, each of the families grows 500 to 600
square meters of vegetables and cereals and they raise small livestock (three to six goats per family) to satisfy the demand of the population not provided for by SONATRACH. But their contribution to the
overall food supply is small. No concrete step toward establishing a
regular vegetable production has been taken. This would certainly
be profitable considering the transport costs from the nearest production areas. The only steps taken in this realm were by agricultural
research centers of the former oil companies, who had been con-
ducting noncommercial experiments for years.

On the other hand, the men (generally illiterate) try to find
temporary jobs with the contractors. In 1972, five interviewed heads
of families worked at contracting businesses for periods of two to nine
months earning an average of DA 600 to 900 per month. The inter-

viewed had been in Hassi Messaoud for several years—often inter-
rupted by periodic migrations. It is their aim to find regular jobs
with SONATRACH. This, however, seems to be rather remote, considering their level of education. However, Castevert and Cote's
seemingly paradoxical statement about the nomads is probably true.
They maintain that these nomads are the most well-balanced and the
deepest rooted in Hassi Messaoud and that they may possibly form
the nucleus of a stable postoil-period population (see also Chapter 7).

Just as in all other oil fields, the food supply of the oil workers
living in residential zones of the oil companies is provided via Algiers.
There are two reasons for this: (1) the Algerian South might not yet
be able to provide the required quality and quantity, and (2) the purchase is centrally directed and financially administered by Algiers.
The local expenses of SONATRACH for food amount to only DA 300,000
per year in the district of Hassi Messaoud. Assuming that the 2,500
workers of SONATRACH (one-third of whom, however, are not always
on the base) each consumes DA 10 worth of food daily, the yearly
expenses just for basic foods total almost DA 6.1 million (see also
Chapter 7).[17] However, approximately DA 10 million worth of small
utensils (household appliances, spare parts, hardware) are purchased
in neighboring towns, benefiting only the trade in these towns.

                              Associated Processes

Until now exclusively based on oil, several industrial processing
facilities have also been built in Hassi Messaoud. The refinery of
the former S.N. REPAL (now SONATRACH) must be mentioned. It
has a capacity of 200,000 tons per year, which, however, cannot be
fully utilized because of a limited production during summer heat.
Production in 1974 was: fuel oil, 75,361 tons; gasoline, 14,117 tons;
butane, 5,343 tons to be bottled; and kerosene, 1,265 tons. The 1973
total output was 114,660 tons. The extension to a capacity of 500,000
or 1 million tons is projected. The production is consumed in the oil
centers themselves and in the Algerian Sahara.

Another facility is the SONATRACH bottled gas filling station,
which fills 20 to 40 cubic meters of butane gas (from the refinery)
per day for household consumption in the south.

Since 1972, SONATRACH's own gas-powered plant in Haoudh
El Hamra (capacity of 22.4 megawatts) provides electricity for their
installations and for Ouargla. Nevertheless, during storms and rains
breakdowns frequently cut off Ouargla from the power supply.[18]

The construction of five gas turbines (with a capacity of 20 to
25 megawatts each) to provide alternating current was contracted in
1975. The total capacity will then be close to 140 megawatts and will

be needed mainly for production purposes.[19] No plans to date are
known to connect the other cities in the Sahara, which up to now are
being supplied by their own diesel generators (Touggourt, El Oued),
to these power plants. An enlargement of, for instance, the refining
capacities or the establishment of petrochemical industries in the
South does not seem to be reasonable because of the high temperatures
that demand additional expenses for cooling systems (June 1 to September 10, 1972: 52 days with a maximum temperature above 40°C,
true maximum above 47°C). In addition, the distances to the actual
consumer centers would cause high transport costs.

On the educational level, a center for training technicians is a
part of the Institut Algérien du pétrole, founded in 1965, with institutes
in Dar El Beida, Boumerdes (near Algiers), and Es Senia (Oran).
The center offers 18-month practical courses acquainting students with
theory and practice of oil-well drilling.

IN AMENAS OIL FIELD

In contrast to Hassi Messaoud, In Amenas is not only located
in the almost precipitation-free region bordering on the Edeien Ubari
sand sea, but also far remote from all permanent settlements. The
nearest are Ghadames in Libya (245 kilometers) and the little oases
Zaouia el Kahla (227 kilometers) and Illizi. The distance by land to
Hassi Messaoud is 614 kilometers (asphalt road). Djanet can be
reached by road (approximately 680 kilometers in length) and like In
Salah, by plane, which makes several flights per week.

Facilities

It was in the Edjeleh field in January 1956 that the drilling DL
101 struck the first oil deposit in the Sahara. Production started in
1960 after the completion of the TRAPSA (Compagnie des Transports
par Pipelines au Sahara) oil pipeline to Skhira in Tunisia (see Chapter
8). This line had an original capacity of 7.5 million tons per year
and has a present capacity of approximately 15.5 million tons per year.
The production of all oil fields in the In Amenas region amounted to
7.18 million tons (in 1971), with Zarzaitine 1.9 million tons, Edjeleh
0.9 million, Ohanet North 0.4 million, Ohanet South 0.3 million,
Tabankort 1.3 million, El Adeb-Larache 0.2 million, and several
smaller fields with a combined total of 1.1 million. This amount of
7.18 million tons in 1971 compared with 12.63 million in 1970. However, in 1971 production attained again the previous level, but detailed
data are not available.

The 30-inch pipeline, leading to Hassi Messaoud via Ohanet and connecting the Illizi basin with the triassic deposits, is only a secondary pipeline without pumps. Its small capacity, however, can be enlarged. These capacities can only be used if the production exceeds the receptivity of the TRAPSA pipeline. With respect to Tunisia, Algeria is obligated to transport 9 million tons—if this level is reached—of the Illizi production fields to Skhira.[20] The latter pipe has a diameter of 24 inches, a length of 775 kilometers, and an overall capacity of 13 million tons annually.

Although only 12 percent of the proved oil reserves and 13 percent of the gas reserves of Algeria are located in the Illizi basin, 63 percent of the possible oil reserves and 70 percent of the gas reserves are assumed to be located there.[21] Because of the increase in world market prices and demands for oil and also to a more profitable production in the South (burdened with high costs in transportation and extras) the production figures will undoubtedly rise.

The structure of the base shows a division into three zones, like all other oil bases: a residential zone (zone résidentielle), an oil processing zone (zone industrielle), and a storage and shipping zone of the processed oil (station de pompage). A map of the residential zone shows the typical functional arrangement of such a base. Various facilities provide diversions for leisure time. The prefab houses (cabins) are arranged in groups according to income levels for workers (eight and four beds) and executives (two beds). The existing 1,160 beds are sufficient, as one-third of the work force are always on vacation. Only the two clubs are housed in solid buildings, all other structures are prefab units.

Labor Force

Because of SONATRACH's assumption of most of the ancillary services (catering, gardening, parts of the transportation and routine repairs) the SONATRACH work force in In Amenas increased to 1,212 persons (January 1973). Of these 8 were executives (cadres), 107 were foremen (maitrise), and 1,081 were specialized and simple workers. A total of 1,433 permanent jobs are scheduled, and an additional 40 foreigners (partly from Arab countries) are in consulting functions. There are always 781 persons at the main base; the others are distributed throughout the various producing fields, returning —distances permitting—to the base at night.

A spot check of all 279 workers in the production sector shows the same predominance of North Algerians in qualified positions (see Table 6.5). These workers were trained during the years of oil prospecting west of the Hoggar Mountains. They then migrated with

TABLE 6.5

Origin and Qualification of the In Amenas Work Force,
Production Division, 1973

| Qualification | North Algerian | | South Algerian | | |
	Abso- lute	Percent of Total	Abso- lute	Percent of Total	Total
Executives	2		1		3
Foremen	26	9.3	4	1.4	30
Workers	71	25.4		62.7	
Total	99	34.7	180	64.1	279

Source: Personal calculations from SONATRACH records.

the oil enterprises to the east after production had started. Their
families, however, always remained in their home villages. Further,
the low representation of South Algerians in the higher income levels
is shown distinctly. This is a result of fewer training opportunities
(also affected by the reluctance to spend the time for possible training
in the North) and to the location of the employment office in Algiers.
It is characteristic that of the 23 interviewed Southerners in Hassi
Messaoud and in In Amenas only one (who had already moved to Algiers with his family) and perhaps another were definitely willing to
settle with their families in Northern Algeria. Several of the interviewed had been to the North once and reported with horror of the cold
and deserts.

As workers are generally recruited from the settled population,
almost no workers come from the closest small settlements of Zaouia
el Kahla (approximately 320 inhabitants, plus 762 nomads) and Illizi
(approximately 650 inhabitants, plus 3,350 nomads).[22]

Ancillary Services and Settlement

Close to the SONATRACH base, a Village des contracteurs is
found with the following enterprises: ALTEST (geological laboratory,
a subsidiary of SONATRACH), two road construction enterprises that
are deserted, hotel-restaurant, gas station, three car and truck repair
shops, four transport enterprises, two of which are private, road
department depot (state), FLOPETROL (pipe cleaning, private), two
private construction enterprises, two private metal construction en-

terprises, various oil drilling enterprises that are deserted. A total
of 500 persons is employed there. Because of the SONATRACH take-
over of numerous contractors, the village today offers a picture of a
chaotic junk yard worse than the Hassi Messaoud center.

In the early 1960s the total population of In Amenas reached
3,600 persons, and is perhaps 2,200 persons today (1966: 2,400).
Only a few of these are complete families. Only five solid family
homes are at the disposal of the SONATRACH workers; another 20
are under construction, and additional 80 are planned. The new
policy to settle families is only carried out hesitantly. Just like in
Hassi Messaoud, usually only members of the higher income groups
(three of the interviewed eight persons with an income of DA 1,800
or more, and five of the total 17) were willing to settle in In Amenas
with their families. Nobody would remain longer than necessary,
which is not surprising considering the climatic conditions.*

To a much greater degree than Hassi Messaoud, In Amenas is
a pure mining settlement, spatially and functionally isolated from the
rest of the Algerian Sahara. However, a certain tertiary sector does
exist: a hospital with 13 beds and a staff of five persons, but without
a doctor (who is available at SONATRACH), is run by the Wilaya administration and serves the residents as well as migrating nomads
who often suffer from tuberculosis.

The postal facility, connected by radio with the North, employs
12 workers, some of whom live there with their families. The public
administration consists of a police station with ten persons, an administration building with an administrateur and 15 employees (In Amenas,
an industrial center autonomous since 1969, is a part of the community
of Djanet), and a customs office.

The shopping center contains three private "drug stores," all
operated by Mozabite merchants, selling canned food, household goods,
stationery, clothing, and medications. Except for goods smuggled
from Libya and the former Spanish Western Sahara, prices are very
high because of high transportation costs. Fresh food such as fruits,
vegetables, and meats is never available. The settlement of a subsidiary of one of the national trade companies (for example, OFLA
for fruits and vegetables) is under consideration, but the problem is
that no surplus food is produced in the entire Daira (county) of Djanet.

The establishment of an agricultural producing cooperative and
the settling of 200 families in Deb-Deb (30 kilometers west of Ghadames, 245 kilometers north of In Amenas) has been planned since May

---

*For the weather station Zaouia el Kahla, the daily average
maximum temperatures are 39–43°C between June and September,
the minimum are 20–25°C.

1973, but will not become reality in the near future. So far, all supplies, at least of the base, are shipped in by refrigerated trucks from Algiers, 1,500 kilometers, or four days, away. No services are required from the region, nor from other Saharan centers.

Associated Processes

No further associated industry or trade processes were observed in In Amenas. In 1974 a refinery with an annual capacity of 300,000 tons was commissioned. It will supply the camp and the extreme southeast of Algeria.[23]

Although the settlement is public and is open to anyone for settling, no "immigration" by those not directly involved or by contractors associated with oil production has occurred.

However, the construction of asphalt roads is important and of permanent significance, as they offer through connections from the north to Ghadames and Djanet (the latter being at the end of an additional 507 kilometers of dirt track).

Also of definite importance are the water drillings that occurred in the course of the oil exploration. The agricultural use of such geologic test drillings in Zelfana (between Ghardaia and Ouargla) is shown in Chapter 4.

The 400-meter deep well of El Adeb Larache (117 kilometers south of In Amenas) produces mineral water of such good quality that the Algerian mineral water company ENA considered installing a bottling plant there. The water could serve as return freight for the food trucks and be distributed in the Algerian Sahara. However, it is questionable whether the extremely unfavorable climatic conditions permit projects of this kind in these regions.

NOTES

1. Annuaire statistique de l'Algérie 1974 (Algiers, 1975).
2. Algeria in Numbers 1962-72 (Algiers: MIC, 1972), p. 8.
3. Annuaire statistique de l'Algérie 1972 (Algiers, 1973), p. 34.
4. "La Wilaya des Oasis, sept ans après le programme spécial," Moudjahid, December 18, 1973.
5. Data from Annuaire statistique de l'Algérie 1974; and Les résultats (partiels) de l'enquête emploi et salaires de 1974 (Algiers: Secretariat d'Etat au Plan, 1975).
6. See SONATRACH information material; B. Murgue, "L'équipement industriel de l'Algérie," Industries et Travaux d'Outremer, no. 267 (February 1976), pp. 84-125; "Signature d'un contrat entre

la Sonatrach et la firme italienne Nuevo–Pignone," Moudjahid, July 1, 1976.

7. Industries et Travaux d'Outremer, April 1976, p. 285.

8. Moudjahid, April 4, 1974; and Pétrole et gaz arabes, no. 141 (February 1, 1975), p. 12.

9. SONATRACH (Algiers, 1972); and SONATRACH information leaflet, 1974.

10. Among others, see C. Castevert and M. Cote, "Mise au point sur Hassi Messaoud," Annales algériennes de Géographie, no. 9 (1970), pp. 106–17; G. Corna Pellegrini, 'Per una geografia delle città pioniere—Hassi Messaoud, pub. no. 4 (Milan: Catholic University of Santo Cuore, 1971); S. Lerat, "Hassi Messaoud," Cahiers d'Outremer, no. 93 (1971), pp. 16–31.

11. According to Corna Pellegrini, Hassi Messaoud, pp. 77 and 81. Unfortunately, more recent figures were not available.

12. See also Castevert and Cote, "Mise au point sur Hassi Messaoud," p. 114. The phenomenon that even less–qualified personnel are recruited from the north, whereas southerners are jobless, is mentioned in "L'équilibre régional," Moudjahid, May 26, 1976.

13. According to RBTOK management in Hassi Messaoud and the owner in Ghardaia; see also Chapter 7.

14. For the history of construction and detailed description, see Corna Pellegrini, Hassi Messaoud, pp. 66–73 and Map 25.

15. Tender was described in Le Monde, August 13, 1974. New data from "Hassi Messaoud—De Nouveaux logements pour les pétroliers," Moudjahid, July 21, 1976.

16. Corna Pellegrini, Hassi Messaoud, pp. 74–77 and pictures 27 and 28; Castevert and Cote, "Mise au point sur Hassi Messaoud," p. 112.

17. This sum seems even to be small in comparison with the information that SONATRACH in Hassi Messaoud has to spend DA 140 per day per oil worker with full board; see Moudjahid, November 5/6, 1972.

18. "Par–ci, par–là," Moudjahid, March 10/11, 1974.

19. Moudjahid, April 9, 1974; Pétrole et gaz arabes, February 1, 1975; and Industries et Travaux d'Outremer, no. 254 (January 1975), p. 63.

20. P. Valberg, "Cinq ans après," Annuaire de l'Afrique du Nord 1970 (Paris, 1971), pp. 51–87.

21. "L'estimation des reserves d'huile et de gaz et orientation des travaux d'exploration ultérieurs au Sahara algérien," Moudjahid, October 7, 1971, p. 4.

22. M. Mahrour, "L'occupation du sol et la distribution de la population dans le Sahara algérien," Annales algériennes de Géographie, special number (Colloque de Ouargla), September 25/26, 1971), pp. 6–14.

23. Murge, "L'équipement industriel de l'Algérie," p. 97.

7

SYNOPSIS: OIL
PRODUCTION AND
REGIONAL DEVELOPMENT

Following this detailed survey, it would appear to be of value
to provide an analytical synopsis comparing results with other regional data. The effects on the infrastructure shall come first.

TRANSPORT AND ADMINISTRATION

The effects of the oil industry on the transport routes and systems are most apparent. In 1949 not a single paved road existed
south of the Ouargla-Ghardaia line, and these cities were connected
to the north by a one-lane asphalt track. Today, a network of two-
lane asphalt roads extends past Hassi Messaoud to Zaouia El Kahla,
El Adeb Larache, Edjeleh, and Ghadames; and in the west from Ghardaia to Adrar and past In Salah to Hassi Krenig. The East Saharan
road network, which eventually will be expanded by a road to El Borma
(Algerian side), has been extended as a direct result of the oil industry.

The railroad from Biskra to Touggourt was changed to normal
gauge in 1958 and still serves for heavy-duty transport. The Second
Four-Year Plan includes a project extending this railroad to Ouargla
and Hassi Messaoud and further on to Ghardaia-Laghouat-Djelfa.[1]

The dense public flight network was possible only because of the
large demand of the oil companies. Hassi Messaoud and Algiers are
connected by flights 18 times a week. In addition, flights to Djanet
and Tamanrasset serve the vacationing workers returning to their
hometowns and villages.

In contrast to the Algerian oil fields, the Tunisian fields have
no effects of this kind. This is partly because of the smaller sizes
of the bases, and partly because the oil is not rated as a regional de-

velopment factor of the same importance as in Algeria (especially
since it is raised by foreign companies). Utilities of the oil settlements such as hospitals, post offices, and administrative departments
are available to all; they supplement the existing structures. The re-
location of the Wilaya Oases Administration from Laghouat to Ouargla
in 1966 was also a result of the shift in the economic center of gravity
toward the oil fields.

If—as projected many times—the oil bases would segregate
themselves from the mother communities and become independent,
the administrative infrastructures would have to be further extended.
This would outlast the oil period and open a basis for future permanent
and stable settlements.

EFFECTS ON THE LABOR MARKET

If we regard the results of our regional research we can state—
as generally done in oil studies—that the oil industry is characterized,
in the Sahara too, by high investments and rather low demands of workers. But this regards neither the small actual labor supply nor the
lack of economic alternatives in the region. In this respect the effects
of the oil industry on the labor market are not inconsiderable and the
monetary effects are of still more importance. They could even be
increased with a more directed involvement of the Sahara population
into the modern sectors of industry and trade.

Traditionally, wages in the oil industry are high, partly because
of the harsh working conditions. In Algeria the average 1973 income
of qualified workers in the oil and natural gas industry of DA 1,020
per month was by far the highest. This compares with the national
average of DA 804 per month paid in the state sector. Because of the
integration of nonqualified personnel from former contracting firms
into SONATRACH, the average qualification level and, accordingly,
average wages of all qualification groups, fell within SONATRACH
from DA 1,417 in 1972 to DA 992 per month in 1973. This figure is
still well above the national average of DA 892 per month in all
branches of nonagricultural activities.[2]

The origin of the workers clearly points out that in competing
for higher positions the local workers are at a disadvantage. In the
case of Hassi Messaoud, for instance, it seems that those jobs made
available because of the departure of the French were filled with workers from North Algeria. A comparison with the work forces of the
contracting firms shows that this is not only a result of the low educational level of the Saharan population, but also, certainly unintentionally,
caused by the centralization of administration headquarters in Algiers.
(A regional personnel employment office in Hassi Messaoud operating
since the fall of 1974 slightly improved the situation.)

SYNOPSIS: OIL PRODUCTION

TABLE 7.1

Origin of the South Algerian Oil Workers in Hassi Messaoud,
1968 and 1973
(spot check)

Region	1968		1973	
	Absolute	Percent	Absolute	Percent
Oued Rhir (Touggourt)	323	40.5	37	35
El Oued	166	21	32	30
Biskra–Sidi Khaled	80	10	14	14
Mzab (Ghardaia)	126	16	16	15
Ouargla	72	9	7	7
Laghouat	14	2	4	4
Golea und Timimoun	10	1	4	4
Total	796	—	104	—

Sources: For 1968, C. Castevert and M. Cote, "Mise au point
sur Hassi Messaoud," Annales algériennes de Géographie, no. 9
(1970), pp. 107–16; 1973 data from SONATRACH records.

For instance, 65 percent of the SONATRACH workers in Hassi R'Mel
come from the north, in contrast to only 30 percent of the 73 Algerian
workers of the ALFOR drilling company. In Hassi Messaoud, approximately 57 percent of the SONATRACH work force are North Algerian;
the Hassi Messaoud private transport company RBTOK employs only
South Algerians.

A distinct zoning within the Sahara is also visible, as shown in
Tables 7.1 and 7.2. This zoning also does not correspond to the
population potential of the individual countries (Dairas) as shown in
Table 7.3.

Moreover, the oil companies preferably employed sedentary,
former agricultural workers and small craftsmen. These workers
were often from the social strata of former black slaves (Harratin)
who are found mainly in Oued Rhir and El Oued. In addition, the
workers already trained in the central Sahara region during the prospecting period of the 1950s were transferred to the extraction camps.
They also consisted mainly of agricultural workers from oases such
as In Salah or El Golea. In contrast, workers from the nomad group
were given work only for seasonal exploration activities, such as
seismic exploration.[3]

An explanation for the relatively few workers originating in the
nearby cities of Ghardaia and Ouargla is the attractive job opportuni-

TABLE 7.2

Origin of the South Algerian Oil Workers in In Amenas, 1973
(spot check)

Region	Absolute	Percent
Golea	48	27
In Salah	47	26
El Oued	21	12
Mzab	20	11
In Ghar	15	8
Oued Rhir	8	4
Aoulef	8	4
Djanet	3	2
Adrar	3	—
Timimoun	3	—
Ouargla	2	—
Reggane	1	—
Tamanrasset	1	—
Total	180	—

Source: Data from SONATRACH, 1973.

ties that were and still are provided in those cities (Wilaya administration in Ouargla, trade in Ghardaia). In Ghardaia, the male work force also prefers jobs in administration and trade—often even in North Algeria—to the hard working conditions in the oil industry.

However, today the native population participates in the activities of oil producing to a much greater extent than is reported in earlier publications. In Saudi Arabia, for instance, of the 18,000 ARAMCO employees in 1957, only 38 percent were from the surrounding vicinity.

The number of foreign workers in SONATRACH (many of whom were leading staff members before the nationalization) has fallen to 1,593 amounting to 4 percent of the 1974 total of 39,663.[4] These were generally consultants outside of the organizational chart of the enterprise. This is seen in Algeria as well as in Tunisia and has not only practical but also economic motivations. The widespread image of oil bases with predominantly foreign work forces, surrounded by mountains of empty beer cans, decorated with pin-ups, and frequently entertained by ballet troupes flown in from Paris is definitely a part

TABLE 7.3

Population in the Dairas of the Algerian Sahara, 1968

Daira	Population	Percent of Nomads	Part of the Total Population in the Sahara (Wilaya Oases and Saoura)
Biskra	203,100	16	22.0
Touggourt	122,600	16	13.2
El Oued	118,970	13	12.9
Ghardaia	83,340	10	9.5
Bechar	66,690	7	7.2
Laghouat	58,230	39	6.3
Adrar	55,290	0	5.9
Ouargla	49,400	43	5.3
Timimoun	40,270	0	4.3
In Salah	25,180	4.5	2.7
Beni Abbes	20,020	8	2.1
Abiod S. Cheikh	17,990	69	2.0
El Golea	16,670	35	1.8
Tamanrasset	16,300	40	1.8
Tindouf	13,150	59	1.4
Djanet	5,890	41	0.6
Total	922,360	17	—

Source: M. Cote, "Sahara—Economie, population, industrie," Annales algériennes de Géographie, no. 9 (1970), p. 140.

of the past. The only reminders of "gold-digger days" are French and Egyptian movies, the large bars on the bases that are gradually being deserted, and the not-often-used swimming pools (if working at all). So, at least socially the oil industry is losing its "foreign body." The oil field is no longer an exhibition piece of Western civilization, but is becoming a normal working place as any other area in the fast-growing, modern sector of North Africa.

Monetary Effects

The overproportionate monetary effects presented in Chapters 5 and 6 are largely a result of the high wages paid in the oil sector. This was augmented by the nationalization in Algeria (wages for qualified workers increased from DA 900 to 1,400 per month).

The attractiveness of the national oil company, especially in comparison to the private contracting firms, is explained by the clear difference in payment and working conditions. In a letter, a worker from Hassi Messaoud quite emotionally told that certain laborers in the construction businesses (contractors) earned only DA 12 per day (while the contractor charged the oil company DA 120 per day for each worker). He also mentioned they paid neither income tax nor social security; that they were not granted the vacation time they were entitled to (three weeks after nine weeks of working); and that their room and board was unsatisfactory.[5] However, since then, personnel of nearly all the contracting firms have been integrated into SONATRACH, raising its effect from DA 22,100 in 1972 to DA 31,061 in 1973 and DA 39,663 in 1974. Laborers and nonskilled workers now earn an average of DA 816 and 580 per month, respectively.[6]

With the otherwise low wages, the multiplying effects of these incomes are important. Each of the persons interviewed by us working in the Algerian oil industry supported an average of 9.6 persons. Also, the interviews clearly showed that in Tunisia as well as Algeria, the relatively low income groups (workers, laborers) spend their wages exclusively in their hometowns for the support of their families and for investments (houses, property). If, in the case of the In Amenas Division Production (279 workers) 15 come from the small oasis of In Ghar (between In Salah and Reggane), it can be assumed that its entire population lives from the money transfers of those men.

Behavioral Changes Within the Labor Force

In spite of all the personal mobility of the oil workers, sometimes traveling more than 1,000 kilometers to their places of work, their residential mobility is not so high. Of the Sahara inhabitants interviewed in Algeria, only two would consider settling in the north and retiring there. Also, no North Algerians were willing to settle with their families in the south. Even company-owned apartments were refused. Of all 31 interviewed, nine (30 percent) had moved away from their birthplace with their families. However, all but one had remained in the region. The functional separation of place of work and place of residence is maintained and is spatially visible just like the separation of the modern working world from the traditional society.

The high income of workers in the oil industry provides them with the opportunity to investment (mainly in their places of origin) in the construction of a house and in property and gardens. Sixty-eight percent of the interviewed in Algeria (average wage, DA 1,530 per month) had already constructed a house or were having one constructed. These figures correspond to the situation in Tunisia (El Borma, 74 percent of the interviewed).

The result of this behavior is a construction boom all over the Sahara and Saharan fringe areas. Adding to this boom are the money transfers from workers in foreign countries. However, only in the rarest cases is the capital invested in facilities of industrial or agricultural production. We cannot decide whether this is a remnant of the nonproductive attitude sometimes associated with the Oriental mentality or whether the public stimuli and the bases for such investments are lacking.

Other studies have already examined the phenomenon of house construction. For instance, the French oil company CFPA in Hassi Messoud conducted random spot checks of 47 workers.[7] Eight of them had constructed or were constructing a house on family property, 13 had a site and had started construction, two purchased an already erected house, and five had not yet built a house but intended to do so. Of the examined dwellings, 23 had electricity and 29 running water. Seventeen were traditionally furnished, although with a Western touch (separate chairs, closets, beds), 16 were in Western style, and 18 were basically traditionally furnished.

Statistical data, although far from complete, show the extraordinary importance of construction in the Wilaya Oases.[8] Whereas on the Algerian average, one apartment per 2,240 inhabitants was constructed in 1972, the ratio in the Wilaya Oases was 1:1,369. Contrary to the other Wilayas, 43 percent of them were privately financed (the average in Algeria was 18 percent). Sari, for instance, in a survey from northwest Algeria, shows that this aspiration for a house is common, once income will allow it.[9] In the mines of Bou Caid (Ouarsenis), an untrained worker earns approximately DA 4,500 per year. With his actual savings (DA 1,700–2,000 per year), he tries more than anything else to solve the problem of living space. Most of all, more space has to be created for the family—at present, rooms of five square meters are lived in by several persons.

Changes in the social behavior are not that easily ascertained. Our survey of house construction habits, at least in Algeria, points to a certain preference of traditional styles in accordance with the requirements of the traditional Islamic family. The spatial separation of family and work place, which is accepted without hesitation, also intensifies the separation of the two living spheres and prevents a transfer of Western ideas (such as women's liberation) to their families.

The educational behavior and conscience is being decidedly affected. All of the interviewed fathers in both countries send their children, including the girls, to school. They frequently hope to or already have sent their boys to high school and/or a university. None of them were willing to let their children learn a traditional profession (farmer, craftsman), although 15 (48 percent) of the Algerians interviewed came from an entirely traditional, agricultural background and another six (19 percent) were traditional craftsmen. None would be willing to return to his traditional job in case of unemployment (if oil no longer provided jobs). They would rather accept a similar, modern job anywhere in Algeria—preferably close to their family residences. Only five (16 percent), three of whom had already worked in France, would be willing to emigrate.

With regard to these facts, a class of modern industrial workers is emerging that by no means can be reintegrated into a traditional economy. However, the development of a new dualism between the well-paid industrial workers in national companies and the workers in the private and traditional sectors has to be avoided. The considerably higher income of the industrial worker affords him the chance to further his and his children's professional advancement, whereas the workers in the private and traditional sectors are impeded in their progress. The contrast between the former nomads and the oil workers in Hassi Messaoud is a striking example.

EFFECTS ON OTHER SECTORS OF THE REGIONAL ECONOMY

These effects can be twofold: First, backward linkages toward already existing structures have positive effects on the latter, and second, as forward linkages, new spatial preferences might be created that stimulate future development.

Agriculture

Several publications deal with the problems of agriculture in the northern fringe of the Sahara, in oases that "import a lot, export almost nothing at all, and produce at high costs."[10] With the breakdown of the labor-intensive Sahara trade and the arrival of new eating habits, the traditional oasis agriculture focusing on crops like dates and cereals no longer made economic sense.

Considering the multitude of informational material, we refer here only to more specialized data and to our own inquiries. The agriculture is still of considerable importance for the Wilaya Oases.

Fifty percent of the working people, or 250,000 persons out of the total population, live from agriculture.[11] The most important product is dates. Currently, 90,000 tons of dates (half of them from Oued Rhir/Touggourt) at a value of DA 50 to 60 million are harvested from approximately 4 million date palms. Other agricultural production, mainly vegetables, has a value of approximately DA 50 million.[12]

The crisis in date cultivation, caused by the decrease in demand and increase in labor costs led to a drop in production (that is, in Oued Rhir, 50,000 tons in 1940 to 40,000 tons in 1971) in spite of an increase in the number of palms. Several measures, such as the central sale of the dates through a national monopoly, the construction of five processing plants in the production centers, attempted to give new vitality to these cultures. It is doubtful that a real, long-term revival through an expansion of the palm groves, a consolidation of the scattered property structures, and a settling of nomads as additional workers will be successful. Nesson shows that even modern integrated date plantations do not have a very bright prospect for an economically viable future.[13]

Upon visiting the oasis Zelfana in May 1973 this problem became distinctly apparent.[14] Indirectly the entire oasis is a result of oil exploration activities because a geologic test drilling in 1948 struck a powerful well (then 31.0 liters per second) at a depth of 1,206 meters. This well was completed by drillings in 1954 (260 liters per second) and in 1968 (80 liters per second).

As early as 1948 and then systematically in 1952 the Caisse d'accession à la propriété et à l'exploitation rurale (CAPER) started cultivation of the date palm plantations, first in the Zelfana region (until 1954), then in the regions of Gouifla (1957–59), Hassi North (1958–60), and Hassi South (1959–62). In 1969 a total of 27,507 palms grew on 330 hectares. Of these, 8,298 were productive, which is said to have meanwhile increased to approximately 20,000. Dimensions and value (estimated) of the total sale of the dates developed as shown in Table 7.4. Two-thirds of the production consists of the exportable Deglet Nour date variety. In addition, the following useful plants are grown on ground level: onions, carrots, potatoes, alfalfa, and bamia vegetables. However, these products are raised only for the local market and are not sold elsewhere.

This means that the total monetary income of the 302 members of the cooperative who, with their families, make up the 2,000 inhabitants of the Oasis, theoretically comes only from the profits from the sale of dates to the national OFLA (Office des fruits et légumes d'Algérie) and private enterprises. Otherwise, they have to subsist on the products (vegetables) that they raise themselves. Subtracting the DA 400 fee for water, land lease, and operating expenses that has to be paid to the cooperative by each member, each producer earns annually DA 2,100 (approximately DA 26 per month and inhabitant).

TABLE 7.4

Development of Date Production in Zelfana, 1969–73

Year	Quantity	Value (DA)
1969/70	293,000	178,730
1970/71	693,000	422,730
1971/72	654,000	398,940
1972/73	1,216,000	735,660

Source: Coopérative agricole, Zelfana.

Therefore, the economic basis of the oasis, which is even subsidized by the government-supported date purchase prizes, is at least questionable. In fact, additional incomes are derived from sporadic jobs with the oil contractors, from money transfers of emigrants, and from wages paid to state officials by the government. The economic problem is not only phenologically visible through deserted houses of settlers and neglected plantations—in contrast to the tents and twig huts of the spontaneously settling nomads—but is also realized by the local authorities. So far, the problem of transportation and the lack of nearby active markets prevents the initiation of a possible market-intensive horticulture. The proximity to the oil-producing centers such as Hassi Messaoud has raised hopes for a market that has not yet been fulfilled. In fact, no products at all are delivered there.

So far, the breeding of livestock serves exclusively the oasis. An expansion is also planned, but after 80 cows had been allocated to the cooperative, there was no use whatsoever for the milk. The economic problems of the oasis are far from being solved, indicated by the fact that only 231 of the 310 distributed plots are cultivated and producing.

Nevertheless, further attempts are being made to develop the traditional agricultural sector by means of new plantations and also by the introduction of market-intensive cultures. For example, after completion of a deep well in 1961 (250 liters per second), 100 poor nomad families have settled in Hassi Ben Abdallah since 1966.[15] Each of the families was assigned to 1.5 hectares of land and 100 palms. A market-intensive cultivation of melons and tomatoes is also under consideration. However, we have to doubt that these projects are economically reasonable. Nearby markets are lacking and distances to other markets, except for the oil camps, are high.

The example of the oasis agriculture in Aoulef and In Salah, which partly changed to the production of winter tomatoes (in 1972/73 the total production of 2,000 tons was flown to Europe) is a particular case.[16] Its economic profitability must be critically examined, too. The sale of products within the region would make more sense, as there are approximately 9,000 workers fed by SONATRACH on the oil bases (one-third of these being on regular or family leave). Therefore, there is a potential annual demand for basic food (including fruits, vegetables, and meat) of approximately $9,000 \times 2/3 \times$ DA 10×365 or DA 22 million, mostly met from North Algeria. This is 20 percent of the value of the current agricultural production (except meat) in the Wilaya Oasis.

It is now necessary to direct the strong potential demand toward regional agriculture. The current attempts to modernize agriculture in the framework of the Algerian agrarian revolution might provide the appropriate means.

Trade and Industry

In contrast to the oil pipelines, whose termini provide preferable locations for industrial development (see Chapters 9 and 10), the oil production itself seems to have a small direct impact on forward- and backward-linked branches of the secondary sector. Except for some facilities in Hassi Messaoud, no processing industries can be observed around oil fields. This is only to a small extent due to the hostility of the natural setting. The most significant reason is the marginal location in relation to the domestic and foreign consumer centers.

However, the indirect effects on the entire economic region of the Wilaya Oases are considerable, especially as all other starting points for an industrial development have so far been lacking. For instance, the actual population growth of the desert town of Ghardaia coincided with the crisis of nomadism and the discovery of hydrocarbons in Hassi R'Mel and Hassi Messaoud. Ghardaia became the corresponding supply center. Thus, population more than doubled since 1954 to 32,000 in 1969.[17] Already in 1960, 22 percent of the population was employed in industrial and small industrial enterprises (construction and transport business, repair and installing businesses, and garages), a part of which is closely connected with the oil industry. The new industrial zone of Ghardaia, established in 1970 by the Algerian Ministry of Industry on the plateau of Garet Taam southeast of the city, is a good example of the multiple interdependence with the oil sector.[18] Decisive for this industrial zone was the government's plan to direct the capital of the Mozabite merchants toward the establishment of a light industry that had been previously neglected by the government.

Until the beginning of government activities in the wholesale sector (1964) in which national monopolies exist in many fields today, Mozabite merchants supplied the entire Algerian south. The main profits were made in this sector and in the transport business during the peak of the oil exploration between 1955 and 1962. Even then they traded almost exclusively with the contracting firms and not with the big producing companies. The decline of the Mozabite Saharan trade began with the increase of state influence in the wholesale sector, the drop in contractor activities, and the growing concentration of all administrative functions in Algiers.

On the 50 hectares of the Ghardaia industrial zone, all enterprises actually existing were financed by the merchants. At present, the following factories have been erected: car radiator manufacturing (RBTOK), underclothes and knitting (MTC), socks (Hadj Bouchanga), plastic bottles, tiles, manufacturing facilities for textile, laces, and rubber bands are under construction.

In addition to a guaranteed water supply by a water storage basin, the most important prerequisite provided by the Ministry of Industry was the construction of an electric power plant equipped with six production groups of 6 megawatts each. The plant started full operation in the winter of 1973/74, after being connected to the Hassi R'Mel-Hassi Messaoud condensate pipeline. It has since then provided electrical power for the industrial zone and the city itself (before, Ghardaia was supplied by Hassi R'Mel).

The provision of a reliable and cheap energy source was an essential requirement for the settling industry (normal customers are charged, like everywhere else in Algeria, DA 0.37 per kilowatt hour, enterprises in the industrial zone are charged DA 0.15 per day or DA 0.08 per night). When compared to the former power supply by individual generators, this resulted in a considerable drop in energy costs. For instance, MTC energy costs dropped from DA 4,000 to DA 1,300 per month whereas SONERAS dropped from DA 3,000 per month to DA 900 per month. The enterprise SONERAS (Société Nouvelle du Radiateur Saharien), belonging to the owner of the transport enterprise RBTOK may be studied a little closer in this respect. It was founded in 1970 by two transport entrepreneurs who raised DA 300,000 exclusively from activities related to the oil sector. Today, the enterprise provides jobs for 55 persons who repair and manufacture car radiators (salaries of DA 400–700 per month). The yearly turnover amounts to DA 2 million. There is no lack of lesser qualified, younger workers, but the situation with specialists is more difficult, as the competition with the oil industry causes a disproportionately high wage level.

Currently, the work force (all men) in the entire zone consists of approximately 300 relatively well-paid workers who are bussed in every day from Ghardaia City. This industrial zone—along with similar projects in nearby Berriane—represents the first attempt to introduce a permanent modern industry into the Sahara independent of raw material. According to the statements of participants, it seems to be successful. This is due to favorable energy and work force factors, even though the raw material sources and the sales markets are located in North Algeria or abroad.

The effects of oil on the regional economy are rather strong because no other activities of economic importance other than those described are conducted in the Algerian Sahara. It is the incipient industrialization that will change this unbalanced orientation as has been seen in the industrial zones of Ghardaia and Berriane. Under government supervision, private capital was invested there by former contracting enterprises doing business with oil companies. In addition, the industry can take advantage of the almost-cost-free energy sources (natural gas) that are now being transformed into electrical power in the new energy centers such as Ghardaia.

So, as a direct result of the oil industry, the economically marginal and passive area of the western Algerian Sahara has become active again after the decline of the trans-Sahara trade. However, it is still too early to make valid predictions about the future.

Settlement

It is hardly possible to quantify in any way the eventual effects of the oil industry on the existing settlements. But it becomes clear that relatively few oil workers come from those cities that developed most intensely in relation to their settlement structure and population (that is, Ghardaia and Ouargla). These two cities have important trade and service functions for the oil fields. Ghardaia's doubling of the population from 20,500 in 1960 to 56,000 in 1973 came parallel with its emerging functions within the oil business.[19] Although its activities currently appear to be turning away from the oil sector and directed toward a more autonomous growth, its development will never be independent from oil. The present Four-Year Plan alone considers the completion of the following projects: a 300-bed hospital with a specialist division; two tourist hotels; a weaving mill; a cultural center; a board of elementary education; the extension of the technical boarding school for 200 students; a large sports field; and a camping ground.

The development in Ouargla is just as complex.[20] There, the population grew from 34,485 in 1955, of which 20,175 (58.5 percent) were nomads, to 48,323 in 1966, of which approximately 6,000 (14 percent) were nomads, and to 55,278 in 1970. Besides the comparatively moderate population growth of 2.4 percent annually, this was caused predominately by the settling of nomadic groups, for which approximately 3,500 new apartments had been built during this time period. Already in 1960, at the peak of the oil boom, Ouargla superseded Laghouat as the seat of the administration of the then Départment Oases.

After the decision on November 24, 1966, to implement a Special Program for the Wilaya Oases, most of the regional authorities (planning, education, law courts, economy, customs) moved from Laghouat to Ouargla. Corresponding administration and residential buildings (for 200 officials) had to be built. Within a city development plan, the new residential, administrative, and industrial quarters were erected south of the old fortress-like center of the city (ksar).

The high demand for labor in the construction business and oil exploration led to a visible decline in oases agriculture. Only today is the latter recuperating slightly as a result of the increase in local consumption and the new residents from the north (administration officials). Without any industry (except for a date-processing center) and without a profitable agriculture, Ouargla is becoming more and more an administration and trade city, a hub of the Sahara traffic competing with Ghardaia. It has become a shopping and day recreational "oasis" for the oil workers. This most recent development is manifested in modern hotels and shopping centers offering luxuries and frequented by families from Hassi Messaoud. The city's present extension is due exclusively to the oil deposits and the government's desire to establish an administration center close to them. Its continuously increasing attractiveness will certainly impede the eventual development of a tertiary sector in Hassi Messaoud.

With regard to these facts, it seems to be questionable whether, for instance, Hassi Messaoud will develop into a true permanent settlement. Corna Pellegrini has delineated generally valid characteristics for "pioneer centers":

Productive functions (of goods or services), especially developed to correspond to economic and political interests from other highly developed countries. These must be strong enough to justify the financial costs and human effort of a settlement in regions that are scarcely or not at all populated and generally unfavorable for living.

An economically active population (predominantly male and young) with a high technical level, strongly fluctuating, and in constant exchange relations with their native regions.

A low level of social integration with the (original) population groups
living in the vicinity, which contrasts to very intense relations
with the population in the mother country.

Purely functional and unattractive urban setting; a ramified road network adjusted to motor vehicles; buildings and an infrastructure
of a temporary and unstable type that are subject to constant changes
along with possible changes of function.[21]

    Hassi Messaoud, for instance, is not yet a true city, but an in-
dustrial satellite. It is already a steady settlement whose structures
are becoming less unstable and whose landscape is achieving a certain
degree of "humanization." The future function could be that of a supply center for the entire Algerian south, but in competition with Ghardaia and Ouargla. It remains a political decision whether an extension leading to a certain dissipation of the settlement pattern would
be desirable. In case of a positive decision, an agrarian sector would
have to be developed in Hassi Messaoud, based at first on the squatter nomads already there. Working there since 1962 is an agricultural
experimental research station supplied by water from the Albian (approximate depth of 1,400 meters). Fruits, vegetables, and fodder
are grown for a complete "farm" with, among other animals, 80
sheep, 5 cows, and 4 horses. The surplus production is usually not
sold on the market but generally given free of charge to SONATRACH
employees.

    In spite of all these possible points of departure, Hassi Mes-
saoud will probably start to depopulate shortly after the eventual end
of the oil production. The theory that no permanent settlement could
develop if it could not achieve additional functions seems to be valid
as witnessed by our visit to the deserted camp of "Maison Rouge"
(between In Amenas and Edjeleh). Built in 1955, about 300 oil men
and contractors were living there in 1959. The move to a new base
at In Amenas with a more favorable location began at the end of 1960
and was completed in 1962/63. Although part of the buildings and installations do still exist, the camp is totally deserted. It is even
avoided by migrating nomads, although many usable things such as
wood and plates could be found there. The plantations have dried up,
and there is no water at the surface (which previously had to be
pumped). The same effect of partial or complete settlement decay after retirement of oil-oriented activities is seen, for instance, in
Iran, especially where no industrial development is viable and agricultural development is limited.[22]

Capital Market

    The effects of the wages paid to the region were already de-
picted above (see Table 6.2). It was also shown that the region partly

benefits from the profits of private regional entrepreneurs working in cooperation with the oil companies. Already during the French period, alternatives allowing the region to benefit from the oil revenues were considered. The fact that the valuable oil left the Sahara without leaving distinct, economic traces, caused some disturbances, especially among the native population. The Organisation commune des Regions sahariennes (OCRS), which, however, spent part of the money for oil-associated investments, and the Caisse de Solidarité saharienne (CSS) had the capital shown in Table 7.5 at their disposal (from oil revenues of the French government). The various organizations succeeding the OCRS were also granted considerable amounts of money.

Combined with the decision in Ouargla on November 24, 1966, to set up a DA 600-million special program for the development of the Wilaya Oases came the statement that DA 15 million annually should be diverted from the oil revenues.[23] They were to be placed at the disposal of the Wilaya Oases. A total of DA 410 million was spent in the framework of this program (which essentially came to an end in 1973) for the following rather general purposes: economic and social improvements in all fields of the local activities; a reorganization and strengthening of the regional administration; and an improvement of the living conditions.[24]

Obviously there were difficulties in investing all of the available capital. Among the reasons why the plans were not always carried out in satisfactory rhythm were time-consuming preliminary studies, difficulties in the supply of equipment and material, and a lacking dynamic spirit in the entrepreneurs. Nevertheless, the foundation was laid down for better infrastructural equipment in the region and for future further development programs. The industrial zone of Ghardaia is a good example of this.

TABLE 7.5

Investment Sums Available from Oil Revenues in the Algerian Sahara
(in million French francs)

	OCRS	CSS
1960	42.3	10.6
1961	90.5	17.5
1962	68.4	16.0*

*Until August 31.

Source: A. Corneille, Les Chaanba (Paris: CNRS, 1968), p. 207.

Physical Setting

The enormous activities of men above and below the earth's surface also affected the surrounding nature. The effects on the changes in the water ecology are of particular importance. Numerous deep drillings struck ground water deposits especially in the Albian formation (at a depth of 1,300–1,600 meters), partially producing enormous quantities of water. The question of a future exhaustion of the supplies has not yet been clearly answered. However, the decrease in the production of the individual wells in the course of the years indicates such a trend.

In contrast, the water in layers close to the surface is already almost exhausted. Exemplary of this is the 1949 Michelin map indicating an "abundance of water at a depth of nine meters" at Hassi Messaoud (happy well), from which the oil settlement derived its name. This well has been dry now for years.

According to statements from regional specialists, even the climate and local vegetation have changed in the Hassi Messaoud region since 1960. Aside from the beginning of spontaneously growing herbs and even shrub vegetation in depressions where the salt water from the oil drillings gathers, a slight vegetation of herbaceous Cram-Cram (which did not exist before) eventually can be found more and more in other depressions. This occurs between Hassi Messaoud and Ouargla. It is assumed that this vegetation is nourished by the nightly dew condensation of water vapor formed by the burning of surplus natural gas.[25] The thin clouds of soot from the same origin that spread out for hundreds of square kilometers in certain atmospheric layers possibly act as a filter and reduce solar radiation.

The meteorological station of Hassi Messaoud reported a precipitation of 8.6 millimeters for 1972, the long-standing average is 10 millimeters since 1960. This is a marked change in comparison to former years with hardly any precipitation.

NOTES

1. See A. Mihoubi, "2ème Plan quadriennal—Les infrastructures économiques," Moudjahid, June 24, 1975.

2. Annuaire statistique de l'Algérie 1974 (Algiers, 1975), p. 212.

3. For the Chaanba-nomads, see A. Corneille, Les Chaanbas (Paris: CNRS, 1968), pp. 203–06.

4. Les resultats (partiels) de l'enquête emploi et salaires de 1974 (Algiers, 1975), table 5.

5. Moudjahid, June 11, 1971, p. 2.

6. Les resultats; and Annuaire statistique 1974, p. 212.

7. "Edute socio-économique sur l'habitat des agents CFPA en Algérie" (Neuilly: EUREQUIP and CFPA, December 1965) (mimeographed).

8. Annuaire statistique de l'Algérie 1974, p. 76.

9. Dj. Sari, "La désorganisation de l'agriculture traditionelle dans l'Ouarsenis, Etudes rurales, no. 47 (Paris, 1972), p. 61.

10. D. Champault, Une oasis du Sahara nord-occidental—Tabelbela (Paris: CNRS, 1969), p. 447.

11. These and the following figures, which partly seem too high in light of our calculations (see Chapter 6) are from "Vers la réorganisation du secteur dattier dans les Oasis," Moudjahid, July 15/16, 1973; see also, with partially differing figures, "La situation économique dans la Wilaya des Oasis," Moudjahid, December 12/13, 1973.

12. Personal calculations for 1970-71 from Annuaire statistique de l'Algérie 1972 (Algiers, 1973), pp. 90–107 and 191–94.

13. C. Nesson, "Un domaine autogéré au Sahara. Le domaine Hamlaoui Brahim à Ourir," Annales algériénnes de Geographie, no. 8 (1969), pp. 149–62.

14. All data according to records of the Community Council and the president of the Cooperative agricole producteur de dattes, Zelfana, Salah Berrouki. See also "Ghardaia au rhythme du développement économique," Moudjahid, June 13, 1973.

15. "Le village du renouveau—triomphe du désert," Moudjahid, December 12, 1973; also M. Rouvillois-Brigol, "Les transformations de l'Oasis de Ouargla," Annales algériennes de Geographie, Special number (Colloque de Ouargla) (Algiers, 1972).

16. See Moudjahid, December 12, 1973.

17. Gerry A. Hale, "City and Sect in the Algerian Sahara," Geographical Review 62, no. 1 (1972): 123–24.

18. All data from SONELGAZ, SONERAS, and MTC plant managers and Board of Trade, Ghardaia, May 1973; see also "Ghardaia au rhythme du développement économique.

19. "Randonné à travers le Mzab," Moudjahid, June 21 and 23, 1973.

20. The following data from Rouvillois-Brigol, "Les transformations de l'Oasis de Ouargla," pp. 45–58; and Les Oasis (Algiers: MIC, 1970), mainly p. 70; also G. Corna Pellegrini, Per una geografia delle città pioniere—Hassi Messaoud, pub. no. 4 (Milan: Catholic University of Santo Cuore, 1971), p. 98; and H. Redmer, "Ouargla-Wandlungen in einer südalgerischen Oasenstadt," Kölner Geographische Arbeiten, Sonderfolge Beiträge zur Länderkunde Afrikas, vol. 5 (Cologne, 1973): 153–76.

21. Corna Pellegrini, Hassi Messaoud, pp. 30–31.

22. Alexander Melamid, "Satellization in Iranian Crude Oil Production," Geographical Review 63, no. 1 (1973): 27–43.

23. Moudjahid, November 5/6, 1972; according to a different source DA 10 million, see "La situation économique des Oasis—Industrie," Moudjahid, December 13, 1973.

24. "La Wilaya des Oasis, sept ans après le programme spécial," Moudjahid, December 18, 1973.

25. S. Lerat, "Hassi Messaoud," Cahiers d'Outremer, no. 93 (1971), p. 19, note 9.

PART

III

**REGIONAL
EFFECTS OF OIL
TRANSPORT AND
PROCESSING**

8

EFFECTS OF OIL
TRANSPORT

Oil and natural gas alone tend to require their own means of transportation, traditional means generally are not used.[1] The switch to standard gauge of the Biskra–Touggourt railroad in 1958, which transported the first production of the Hassi Messaoud oil fields to the coast, is an isolated example. On the whole, no special impact on the usual infrastructural facilities—aside from some road construction—can be confirmed. The typical and common means of transportation for oil and natural gas is the pipeline. However, as it only transports oil or gas, the pipeline is of no use to other fields of the regional economy.

The oil transport routes (presented in frontispiece) themselves are almost exclusively of phenological significance. Pipelines and tank farms require great spaces, adding visible and often morphological elements to the landscape. In contrast, their direct importance on the regional economy is trifling. These activities require, of course, high capital investments and a certain number of workers during the construction period, but in operation they need only an extremely small personnel for maintenance work. For example, in 1969 the former SOPEG oil pipeline Haoudh El Hamra/Hassi Messaoud–Bejaia (600 kilometers in length, 24 inches in diameter, a yearly capacity of 17.5 million tons) required an operating work force of only 678 persons on its entire length (including the shipment facilities in Bejaia—see later in this chapter). Until that time the capital invested in the installments amounted to DA 488 million. This means a ratio of DA 0.72 million per employee. For the TRAPSA (Compagnie des Transports par Pipeline au Sahara) pipeline, the investments actually seem to lay at at least DA 1.5 million per employee. For the Hassi Messaoud–Skikda pipeline, SONATRACH invested DA 781 million, or DA 1.61 million per employee.

As will be shown in the following chapters, subsequent effects on the labor market are thus minimal. The significance is on a completely different level. An extremely cheap energy source is available at all places where these transport lines end or can be tapped for a specific demand. If this advantage is augmented by a favorable transport location (such as is the case in many oil shipment harbors) it becomes an ideal place for domestic and world-market-directed industry based on cheap energy and/or hydrocarbons as raw materials. Effects yielding and reinforcing self-sustaining economic growth are active in these terminals as is shown also by examples from abroad. The creative power of mineral-oil long-distance pipelines materializes in places where they discharge their oil. There, visible effects are the implantation of refineries and a great number of related industries that strongly influence the regional structures.

The innovational impact can only become effective at these terminals of pipelines when (1) additional positive locational factors exist, such as labor, transport facilities, capital resources, and (2) when these developments are promoted by corresponding national planning. Recognizing these factors early, the Algerian planners marked out these terminals as eventual growth poles for heavy industries.[2]

Using the oil harbors of Skhira, Bejaia, Arzew, Skikda, and Gabes as examples, an attempt to examine the decisive factors and the actual development will be made.

THE TRAPSA HARBOR IN SKHIRA, TUNISIA

Skhira* is located in the southern part of the Sfax Gouvernorate. It is in an area of transition between the Sahel of Sfax and the littoral region of the Gulf of Gabes.[3] In spite of the low precipitation (200–300 millimeters per year) Skhira can only partly be called a part of the arid south of Tunisia as the cultivation of olives and cereals is still possible without irrigation.

Location and Facilities

The following pipelines end in Skhira: the pipeline from Zarzaitine (near In Amenas), 775 kilometers in length, 24 inches in diameter; the pipeline from Douleb, 165 kilometers in length, 6–8 inches in dia-

---

*Also la Skhirra or Sekhira (Arabic: al shīra, probably from classical Arabic: Ṣuḥūra, which means the rocks).

meter; and the pipeline from Sidi El Itayem, 81 kilometers in length, 8 inches in diameter. Approximately 8 to 10 million tons of oil are shipped out of the harbor annually. The capital of TRAPSA is 100 percent French and held by ELF-ERAP.

With only one pumping station, the original and yet most important pipeline from In Amenas had a capacity of 7.5 million tons per year, which, with the installation of three additional pumping stations in 1964, increased to 13 million tons per year, and to 15.5 million tons per year in 1970 (after reinforcing the pipe on Tunisian territory). Until that time, approximately 675 million French francs were invested in the entire facility.[4]

The Algerian part of the pipeline, including the pumping stations SP1 and SP2, was nationalized on February 2, 1971, and has since been under SONATRACH management. This part, however, will not be considered in the following discussion.

The harbor was constructed under an agreement with the Tunisian government dated June 30, 1958; the first crude oil was shipped on September 10, 1960. It is under the company's management and serves exclusively for oil shipment. The fenced-in area contains, besides tanks and technical facilities, administration buildings, a catering complex, sporting and recreation facilities, and 15 villas for leading executives, some of whom are Europeans. The entire area is closed to the public—for security reasons. No links are provided between the ships (including their crews) and the regional economy. Because of the short stays (up to 20 hours), shore leave is rare and the harbor does not have any supply functions for the ships.

Almost no TRAPSA collaborators have settled down in the area of the installation because the harbor has no connections with the region and is not open to the public, and also because two attractive cities—Sfax[5] with 250,000 inhabitants and Gabes—are at a distance of only 75 kilometers and 55 kilometers, respectively, away from the harbor.

Labor Market in Sfax Governorate

On January 1, 1973, the Sfax Governorate had an estimated population of 497,013 inhabitants,[6] 273,270 of whom lived in the Sfax Delegation (city plus suburbs) and 13,817 in the Skhira Delegation. In 1973, 98,000 (19.7 percent) of the population were economically active as shown in Table 8.1.

On the whole, a certain industrial activity (27.7 percent of all employees in industry, 5.5 percent of the total population) is exclusively concentrated in Sfax. There are phosphate (in 1966, approximately 1,200 employees) and food-processing industries (olive-oil

OIL AND REGIONAL DEVELOPMENT

TABLE 8.1

Economically Active Persons in Sfax in 1973,
According to Branches

	Absolute	Percent
Agriculture and fishing	47,000	48.0
Mining	800	0.8
Industry	27,200	27.7
Building trade	3,000	3.0
Electricity and water supply	500	0.5
Trade, banking, Insurance	12,500	12.9
Transport and communication	4,500	4.6
Services	2,500	2.5
Total	98,000	100.0

Source: Governor of Sfax, 1974.

factories, with 2,168 jobs); textile industry (1,200 jobs); lumber (1,093) and metal processing industries; as well as leather and shoe factories (1,275 jobs).

Data could not be obtained concerning the unemployment or the actual demand for a potential work force. The number of emigrants who left the governorate between 1968 and 1973 (7,256 persons, 59 percent of which went to Libya) perhaps indicates that the problem of unemployment has yet to be solved.

With regard to the Tunisian average (25.9 percent of all persons are employed, 28.1 percent economically active) these employment figures appear to be low. Considering the same ratio of economically active men and women in Tunisia overall, the 98,000 job offerings would be met by 139,660 demands resulting in 41,660 potentially unemployed.

Effects on the Labor Market[7]

In 1973, TRAPSA employed 272 persons, of which 19 (7 percent) were foreigners. The number of employees is decreasing; in 1967 283 persons were employed in the Tunisian part, of which 41 (14.5 percent) were foreigners. In the meantime, the latter have been replaced mostly by Tunisians; a process that is not only politically desired for the host country, but is also economically meaningful to the company. This is attributed to the fact that the per capita wage expenses (taxes included) for the locals currently amounts to an average of TD 130 per month, whereas for foreign workers they run up to TD 653.5 per month. This is not only a result of the higher qualifications of the foreigners but also a result of the high "extras" associated with their employment (separation compensations or family allowances, paid vacations, more expensive accommodations). Table 8.2 shows the persons who are constantly employed in the framework of contracts with regional enterprises (calculated on a whole year base) who join the 272 direct employees.

Another TD 46,000 per month are spent for additional transport services carried out by regional enterprises (trucking and transport of the workers from Sfax to Skhira). Several buses and trucks operate exclusively for TRAPSA, adding approximately another 30 persons to the employed. Taking this into consideration, the total number of workers active within TRAPSA is 19 foreigners plus 476 workers from the region (253 Tunisian employees (of which 241 are residents of the Sfax Governorate and 223 contractors all living in this governorate). This amounts to only 0.5 percent and thus represents only a small part of all the employed in the Sfax Governorate.

TABLE 8.2

TRAPSA Contractors in Skhira

	Number
Tugboat service of the Union des Remorqueurs de l'Océan	40[a]
Longshoremen	20
Road and civil engineering (Kilani firm)	45
Catering	50
Gardeners	14
Painters	10
Air transport	3
Greasing entrepreneur	1
Housemaids	15
Total	198[b]

[a]Including 5 foreigners.
[b]Including 193 Tunisians.

Source: TRAPSA management data.

Monetary Flows

Wages paid amounted to TD 546,000 in 1972, TD 397,000 going to Tunisian workers. The total expenses of the company for its own personnel and contractors amount to approximately TD 900,000 per year. According to information of the company, the average income of a Tunisian TRAPSA employee comes to approximately TD 1,459 per year of TD 122 per month. According to our survey, the average net income (taxes deducted) of Tunisian workers is about TD 100 per month and for persons employed by the contractors it is about TD 27. This makes the following wage total available:

TRAPSA without foreigners	$253 \times$ TD 100 = 25,300
Contractors without foreigners	$223 \times$ TD 27 = 6,021
Total	31,321

As much as 95 percent of the TRAPSA employees live and spend their income in the Sfax Governorate. All of the contractors come from the vicinity. According to this, the total of the wages paid and spent locally comes to

$$95\% \text{ of TD } 25,300 = \text{TD } 24,035$$
$$\text{TD } 6,021$$
$$\text{TD } 30,056$$

Of the governorate's total population of 493,000 in 1973, 98,000 were employed. Assuming a monthly average of TD 25 per employee (which is, without doubt, lower than the monthly income of the TRAPSA contractors), the total sum of the wages paid within the Sfax Governorate approaches TD 2.450 million. The wages paid by TRAPSA make up 1.22 percent of this sum, which is a much higher share than that of the absolute number of employees. Still, TRAPSA has only a small significance for the labor market in the governorate.

Economic and Social Behavior
of the Labor Force

The regional and social behavior of the workers in the TRAPSA installations were roughly indicated by 17 interviews (two with women) (see Table 8.3). These interviews yielded surprising parallels to the results in the oil fields.

The two interviewed women (secretaries) were between 20 and 30 years of age, single, and earned (for Tunisian standards) high wages of TD 70 and 80 per month. Both live with their parents and

TABLE 8.3

Origin and Behavior of the TRAPSA Employees in Skhira

Average age of the interviewed	39	years
Average number of supported family members	7.7	persons
Origin from traditional milieu	7	men
New house under construction or built/purchased	14	men
Read to emigrate if necessary	5	men
Native village as projected place of retirement	13	men
Wish to send their children to school/high school	15	men
Prepared to return to traditional milieu	1	man

Source: Personal interviews, May 1973.

aspire to emancipation in the Western sense. This, however, can be achieved only at the present in foreign countries, not in Tunisia.

The analysis of the interviews with the workers shows that

an occupational estrangement or even alienation from the traditional
  background exists;
deep roots in their native village decrease within higher income
  brackets;
there is a readiness to spend and invest their income locally (construction of a house, purchase of fields and gardens for tenants);
there is no will to emigrate (or only as an ultimate solution);
there is a new awareness of the necessity for high-quality training
  for their children (girls included);
a high number of family members may be supported by the relatively
  high income.

As the sole location of modern industry in the region of the governorate south of Sfax, the TRAPSA harbor through its work force has a multiplying effect regardless of numerical restrictions. This will be only partially visible in the delegation and in Gabes, but will be more so in Sfax. In these two cities, however, the influence is not as readily quantifiable as it is in the Skhira Delegation, which shall be used as an example in the following.

Effects on the Skhira Delegation

The modest settlement of Skhira village—since 1968 seat of a delegation (lowest administrative unit)—developed near the existing

railroad station at the intersection of the main road and the track to
the small fishing harbor of Old Skhira (not permanently inhabited).
This took place after the construction of the TRAPSA harbor in 1960.
There are no visible functional relations—not even a direct road link
—between the oil harbor and the village, which is located 10 kilometers
away. Therefore it might be assumed that the strong growth of the
settlement, which is immediately apparent, owes its existence to
other reasons (such as favorable location at railroad and transit road;
absorption of a concentration process of rural settling activities by
the establishment of a supply sector; medical care station and physi-
cian; trade; bus station; and promotion of construction activities by
the urbanization plan). This urbanization plan, which has been roughly
carried out since approximately 1966, shows an orientation of the settlements to both traffic links, namely harbor (Old Skhira)-railroad
station and long distance road GP 1 (Tunis-Sfax-Gabes-Medenine).
No industrial activities can be observed aside from several retail
stores and the slaughter house (for cattle raised by still half-nomadic
groups in the surrounding steppe).

A construction boom started after the completion of the TRAPSA
complex in 1958-60. This provided work for up to 1,000 laborers with
horse and donkey carts within the framework of the aforementioned
urbanization plan. The following structures were built (among others):
104 new apartments (urban and rural type, 1963-71), a National Guard
post (1964), a bus terminal (1969), the seat of the delegation and executives' apartments (1969), an infirmary (1969), a post office (1969),
a storage house and an office of the national cereal supply company
(1969), a lumberyard and office (1969), a community center (Maison
du Peuple) (1972), a mosque (1972), a Moorish bath (1972), and a
sporting ground (1973). [8]

The settlement that developed has the aspects of a small town
with comparatively complete facilities. It is therefore attractive for
the people in the surrounding rural area, where settling and concentration processes are the predominant social and spatial processes.
However, the strong attraction of Sfax prevented the development of
genuine and concurrent urban centers within a radius of 60 to 80
kilometers around the latter town.

This and the traditionally wide dispersion of rural settlements
made the provision of infrastructural facilities a difficult task. This
was to be achieved by the creation of "regional development units,"
which were to provide social and administrative infrastructures. The
construction of 2,000 housing units per development unit was also
projected. Skhira was to become one of these development units.
However, the plans in the meantime have all but failed. [9] The concentration impact triggered by Skhira, however, seems to be continuing.

Since 1966, the statistics of the population growth in the individual delegations of the Sfax Governorate as well as in the three sec-

tors of the Skhira Delegation (Skhira, Rouibta, Sbih) have been extrapolated. This extrapolation presumed a population increase of
13.3 percent until 1973. Thus, special dynamics of the population
growth in Skhira in particular cannot be shown. However, the development of settlements is so remarkable in view of the absence of
other positive factors (besides the favorable traffic location) that the
impact of TRAPSA must be assumed.

In fact, of the 5,355 inhabitants (1973, extrapolated from 1966),
5 are employed by TRAPSA and 74 by the various contractors (mostly
by the Kilani civil engineering enterprise). This amounts to 1.4 percent of the inhabitants or 7.5 percent of the economically active population (if we assume that, according to the Sfax Governorate average,
19.7 percent of the population or 1,055 persons are economically active). The net available incomes of these people appraoch 5×100
plus $74 \times$ TD 25 (because of many day laborers) equaling TD 2,350
per month. This is 10 percent of all of the total wage incomes (cal-
culating the total sum of the monthly available wages at $976 \times 22* =$
$21,472 + 2,350 =$ TD 23,822).

Thus, it is demonstrated here that the effects of oil installments,
on a microregional level, can be felt but are not really relevant. The
reasons for this are the small demand for labor, the trend to segre-
gation from the other sectors of the regional economy, and the exclusive intertwining of the oil sector with foreign markets. Although the
demand is numerically small, the workers have to be qualified and
can only be found in greater cities. Thus 83 percent of the TRAPSA
workers live in Sfax, 8 percent in Gabes, and only 8 percent in the
Skhira Delegation.

Eventually, associated oil-processing industries would have had
much greater regional effects than the oil shipping facilities. But an
industrial enterprise has neither settled there nor was it desired by
the harbor company. Furthermore—except for the available energy
and the favorable transport facilities—all other prerequisites for an
industrialization, which would have to compete with Sfax and Gabes,
are missing. Skhira had once been considered as the location for
Tunisia's projected second refinery, but Gabes (see Chapter 9) was
finally selected.

---

*This number has to be smaller than in the entire Sfax Governor-
ate, which includes the relatively high-paid workers of the city of
Sfax.

Infrastructures

The pipeline as a transport facility is not at all connected with the transportation demand outside the oil sector. Therefore, an improvement of the infrastructural utilities cannot be observed, except for the laying-out of some tracks. A Piste de Service parallels the TRAPSA pipeline; however, its condition is so poor that regular traffic—even with cross-country vehicles—is out of the question. The same is true for the air fields located at fixed distances along the pipeline, as they only serve internal traffic. However, because of their water wells, these pumping stations situated in the desert have a certain attractiveness to the nomads living in that region. For instance, a spontaneous settlement of Bedouins (8 to 12 families) developed at station SP2 (in the region of Segaf, south of Ghadames). Nomads also often stay at station SP4 (El Kamour, west of Bordj Bourguiba) for longer periods of time. Station SP3 at the well of Tiaret (west of Sinauen) houses a TRAPSA personnel of 27-30 persons who are supplied from Sfax.

By no means can the regional relevance of this pipeline be compared with that of the bigger and older pipelines in the Near East. In connection with water drillings, the oil companies operating in those countries report that the news of a water drilling (29 drillings altogether) along the 1,707-kilometer TAP-line to Sidon spread like wildfire. From all directions thousands of Bedouins came with their flocks. In this case, the four big pumping stations did not mean anything else but the construction of four modern homesteads in the middle of the desert.[10]

The technologic progress eventually replaced the manual safety checks of the pipeline with electronic devices and electric control. This was one of the reasons for a decrease in demands on the labor market (at a capacity of 20 million tons of crude oil per year, the Sidon TAP-line in 1955 provided work for as many as 1,100 Arabs). Also, the pipelines' dependence on transport infrastructures, which might eventually have been useful to the regional economy, became less, too. Therefore, except at its terminals, the pipeline is at most a phenological element of the landscape (whether or not visible at all on the surface).

BEJAIA, ALGERIA

In Algeria, the city of Bejaia is an example of a "classical" oil shipping harbor, which until now has no concomitant activities with forward or backward linkages to the pipeline.[11] SOPEG (Société petrolière de Gérance, founded in 1957 by the French companies

CFPA and S.N. REPAL) built the Haoudh el Hamra (Hassi Messaoud)-
Bejaia pipeline (24 inches in diameter and an actual capacity of 17.5
million tons of crude oil per year). The first tanker was filled on December 1, 1959. In accordance with the interests of its French owners,
SOPEG dealt exclusively in the handling of the oil. The transfer from
the pipeline to the ships takes place in a part of the harbor (closed to
the public) constructed solely for this purpose.

At the end of the 1960s, Bejaia was Algeria's largest exporting
harbor (1968: 15.43 million tons of crude oil plus 0.126 million tons
of other export items; 0.165 million tons of import items). In 1973,
Bejaia was only third after Arzew (19.3 million tons of crude oil)
and Skikda (12.6 million tons). It exported goods, including oil, at
a total of 10,387,000 tons, while the Haoudh el Hamra-Bejaia pipeline
alone brought 12.24 million tons. The difference between the two figures is not explained. [12]

The cited studies, however, show that the amount of goods
transited has not affected the region in many ways. The oil that
passes through Bejaia brings about neither an economic reorientation
nor an economic profit. The influence on the urban labor market is
low. Job offerings in Bejaia covered only 53 percent of the potential
demand (all men of working age) in 1966. Table 8.4 shows that only
4.5 percent of the actual offerings emanate from the oil sector.

Along the entire length of the pipeline, in 1968 the company employed 678 workers, of which 206 were French. In Bejaia proper,
358 workers were employed, 93 of whom were French. Since the
wages paid by SOPEG by far exceeded the wages generally paid in the
city, SOPEG paid approximately 13 percent of the incomes available
in the town (1969).

In 1971 the SOPEG company was nationalized and taken over by
the Algerian SONATRACH. At that time, directives for Algerian industrialization were already fixed. Thus the Algerian state industrial
planners had chosen the terminals that were entirely or partly under
Algerian control to be developed as locations for future industrialization activities. This put Bejaia at a disadvantage and, in contrast
to Arzew and Skikda, until 1974 no facilities were planned that would
utilize oil or natural gas as raw material and/or energy supply.

The industrialization of the city (which had grown to 90,000 inhabitants by 1974) had been based exclusively on the raw materials
from the hinterland. The almost 2,000 jobs created between 1971 and
1974 were provided by the wood, cork, and garment industries. The
additional 4,500 jobs to be created in accordance with the 1974-77 plan
are also not related to the oil/energy sector. [13]

Once a location has been chosen for an oil-processing industry,
a corresponding self-sustaining concentration process brings about a
dynamic improvement of locational qualities. It thus surpasses other,

TABLE 8.4

Employment Provided in Bejaia, 1966

Sector	Algerian Work Force Men	Women	For- eigners	Total Sector	Percent of Total
Fishing industry	73	0	0	73	1.03
Agriculture and forestry	251	0	4	255	3.62
Gas, electricity, water	76	0	1	77	1.09
Oil	213	0	104	317	4.48
Industries and crafts	764	1	27	792	11.18
Textile industry	349	10	24	383	5.42
Construction indus- try	704	0	12	716	10.12
Harbor (without oil harbor)	225	0	20	245	5.78
Transportation	406	0	3	409	3.46
Trade	1,193	7	28	1,228	17.37
Banks, insurance companies	18	1	1	20	0.28
Entertainment	26	1	0	27	0.38
Private services	402	157	6	565	7.98
Liberal professions	39	0	6	45	0.63
Administration	993	179	211	1,383	19.57
Military	113	0	0	113	1.61
Small crafts	117	0	0	117	1.65
Undetermined	306	0	0	306	4.35
Total	6,268	356	447	7,071	100

Source: D. Dj. Amrane, "L'emploi a Bejaiar" (Algiers: University of Algiers, mimeographed manuscript, 1970), with figures from the 1966 census.

originally equivalent, locations. This is precisely the reason why Bejaia is now rather marginal in the actual industrialization process of Algeria.

It was not before the end of 1974 that SONATRACH launched a project to construct another refinery in Bejaia. In 1975 British and French companies were contracted to complete it by 1979. At a cost of DA 600 million, the refinery will have a capacity of 7.5 million tons annually and will create 500 new jobs.[14]

NOTES

1.   The North African oil and pipeline system and its details shall not be discussed further, because it is mentioned in the regional chapters. See, for Tunisia, Chapter 2, for Algeria see various SONATRACH presentations; also K. Sutton, "L'industrie algérienne du gaz et du pétrole—Développements depuis l'indépendance," L'Information Géographique (Paris) 34, no. 5 (1970); 212-18; and H. Mefti, "Le transport par pipelines en Algérie," Revue algérienne des Sciences juridiques, économiques et politiques (Algiers) 8, no. 2 (1971): 479-90.

2.   Mefti, "Le transport par pipelines en Algérie," p. 482.

3.   H. Mensching, Tunesien, Wissenschaftliche Länder Kunden, vol. 1 (Darmstadt, 1968), maps on pp. 175, 237, 238.

4.   According to Mefti, "Le transport par pipelines en Algérie," pp. 480-81.

5.   For the town of Sfax, see also the study of urban planning, Sfax 72—Problèmes perspectives d'aménagement (Tunis: Direction de l'Amenagement du Territoire, 1974).

6.   The following numbers are official extrapolations of the 1966 census. From correspondence of the Governor of Sfax Governorate, 1974.

7.   All figures in this section are according to information and records of the administrations in Tunis and Skhira and personal inquiries.

8.   According to a list of the administration of the Sfax Governorate at the end of 1973.

9.   According to M. Fakhfakh, "L'influence d'une grande ville sur l'habitat de sa région: Sfax," Maghreb et Sahara—Etudes géographiques offertes à Jean Despois (Paris, 1973), pp. 137-46.

10.   From ARAMCO reports.

11.   Most of the data in this section are from D. Dj. Amrane, "L'emploi à Bejaia" (University of Algiers, mimeographed manuscript, 1970).

12.   Annuaire statistique de l'Algérie 1974 (Algiers, 1975), pp. 145 and 174.

13.   For instance, see "3,500 logements prévus à Béjaia," Moudjahid, June 9, 1974.

14.   Petroleum Economist, September 1975, p. 345; and Industries et Travaux d'Outremer, April 1976, p. 285.

9

REGIONAL EFFECTS OF
OIL-PROCESSING
INDUSTRIES IN TUNISIA

The case studies in Chapter 8 referred to the terminals of oil pipelines and to oil harbors as not being integrated into the national and regional economy. The following pages show that only the planning concepts of the producer countries can lead to such integration making use of the favorable locations of the oil terminals. We begin with the town of Gabes, where the industrial development is only partly due to its location at the terminal of the natural gas pipeline from El Borma. This demonstrates well that these potentials often enter late into the awareness of the planners. This discussion of Gabes supplements information about the economic development in the Tunisian south (see Chapter 5).

GABES AS AN INDUSTRIAL GROWTH POLE

The Tunisian south consists of the city of Gabes (approximately 40,000 inhabitants in 1966, another 24,000 in the urbanized vicinity; in 1973, it was calculated to approach 47,000 for the city proper) and its governorate (1966: 203,713 inhabitants, of which 111,195 or 54.6 percent lived in rural regions; in 1975 approximately 222,713 inhabitants) as well as the Medenine Governorate. It has been a geographically marginal and, so far, economically passive region. The town of Gabes lies in the climatic zone of the Sahara. A reliable agriculture in this oasis (average precipitation of 190 millimeters per year) is usually possible only with irrigation.

Economic Problems in the Gabes Governorate

At the time when Tunisia became independent, the regional economy was unbalanced: agriculture and the raising of livestock were backward and lacking capital, crafts were decaying, and industries did not exist. General underemployment spread after the departure of the French garrison (which at times consisted of as many as 20,000 soldiers).[1]

Between 1956 and 1966 the population increased relatively slowly at an annual rate of 1.6 percent (17.6 percent altogether), compared with a Tunisian average of 2.72/31.7 percent. Except for Gabes, the few existing modern settlements that could provide starting points for a concentration of population have so far not done so to the same degree as Medenine. Only El Hamma, Douz, Kebili, and Oudref are characterized as "agglomerations" to which, today, Mareth might also be added. The population in the Northern Delegations (Gabes, Mareth) grew more than average, whereas the growth in El Hamma, Douz, and Kebili each achieved only approximately 54 percent of the average Tunisian increase.[2] In the individual delegations, the residential population changed as shown in Table 9.1 between 1956 and 1966.

The immigration losses, which had reached 1.2 percent yearly between 1956 and 1966, continued at an annual rate of 0.6 percent between 1966 and 1971. The income structure is just as unfavorable as in the Medenine Governorate (see Chapter 5). Of the 39,569 economically active inhabitants (19.4 percent of the population; Tunisian aver-

TABLE 9.1

Changes in the Residential Population in the Gabes Delegations,
1956–66

Delegation	1956	1966	Percent of 1956 Total	In Percent Yearly Total	Rural Population
Gabes	60,279	75,228	24.80	2.20	1.80
El Hamma	27,927	32,251	15.50	1.42	3.43
Mareth	12,254	16,577	35.30	3.10	—
Douz	17,429	20,343	16.70	1.52	2.21
Kebili	29,928	33,267	11.20	1.45	2.15
Matmata	25,256	25,924	2.60	0.23	—

Source: Les villes en Tunisie (Tunis: Direction de l'Aménagement du Territoire, 1971), p. 459 (improved).

age: 21.5 percent), 54.7 percent (Tunisian average: 45.8 percent)
were working in agriculture and only 15.6 percent in the industrial
sector. Assuming an overall Tunisian rate of activity (potential work
force, men and women) of 28.1 percent, in 1966 17,637 persons would
have been unemployed as a result of 57,206 job demands and only
39,569 job offers. Even in the most favorable cases when the female
work force is not considered, which would bring the activity rate down
to 21.8 percent (only men), still approximately 4,900 persons would
have been unemployed (not counting the underemployed in the agricultural sector).

In the meantime, the natural growth of the population has further
aggravated the pressure on the labor market. At an overall rate in
the increase of the Tunisian population of 19.3 percent between 1966
and 1972, the population would have grown to 242,879 persons minus
14,700 emigrants to a total of 228,173 persons by the end of 1972.
Since the possibilities of developing the agricultural sector are extremely restricted by limited ground water potential, these jobs
will have to be provided by the industrial sector (if we do not consider
emigration as the only solution).

The importance of this task for the town of Gabes and its near
vicinity is shown in a 1972 survey. The potentially active population
belongs to the fields of activity shown in Table 9.2.

The Tunisian state has long recognized the problems of the
south. In the Perspective Décennales de Développement, published
in 1962, Gabes was mentioned as a possible "developmental pole" in
the Tunisian south. In 1963 these considerations materialized, and
in the Four-Year Plan 1965-68, the city was defined as a developmental
pole (besides Menzel Bourguiba) with a concentration of chemical industries. At that time, the only concise project (with an eventual
investment sum of TD 20 million) was the ICM (Industries Chimiques

TABLE 9.2

Active Population and Employment in Gabes, 1972
(in percent of total population)

Region	With Permanent Employment	Without Employment	Marginal and Underemployed
Gabes City	5.2	20.4	12.6
Gabes and vicinity	5.6	26.2	13.4

Source: Gabes 73—D'un centre agricole à un pôle de développement (Tunis: Direction de l'aménagement du territoire, 1973).

Maghrébines) plant.[3] Construction of the ICM plant started in 1969.
The company was founded in 1962 with public (approximately 50 percent), domestic, and French capital. The production of concentrated
phosphoric acid (54 percent P_2O_5) started on February 18, 1972; and
on May 10, 1972, the plant was officially inaugurated by the president
of the state. Until that time, construction costs amounted to TD 11
million.

On the basis of imported sulfur and phosphates from Gafsa
(consumption approximately 500,000 tons annually), 120,000 tons of
phosphoric acid at a value of approximately TD 6.3 million are projected to be produced annually. Besides 250 permanent employees,
another 1,200 persons are indirectly employed.[4]

This short survey shows that the energy resources of the Tunisian south were not relevant to the original concept of a developmental pole. Much more decisive were regional-political considerations
with regard to the situation of the labor market and the favorable
traffic location. However, the energy deposits (discovered in 1964)
not only accelerated the ICM project but the planning of further installations as well. The host of projects will turn Gabes into an important industrial center in the near future, its region is therefore dealt
with in detail.

Industrial Facilities in Gabes

For security reasons and because Gabes in 1958 did not have a
harbor, Skhira was selected as the terminal of the TRAPSA pipelines.
Now, the cheap energy is supplied by the natural gas pipeline from El
Borma to Gabes-Ghannouche. The oil from El Borma contains approximately 300 to 400 cubic meters of natural gas per cubic meter. This
corresponded to a production of 1.12 billion cubic meters of gas in
1973. The gas had previously been burnt as there had been neither
industries nor other consumers in Gabes.

Not until 1969 did the national Société tunisienne d'Electricité
et de Gaz (STEG) contract a French engineering firm to make a feasibility study on how to use the gas in future industrial plants in Gabes.
Consequently, the construction of the 300-kilometer pipeline with an
annual capacity of 300 million cubic meters was ordered. Operation
began in November 1972.[5] The construction costs, partly covered by
loans from Kuwait, ran up to TD 8 million, maintenance costs will
amount to approximately TD 0.3 million per year.[6] In 1976 the capacity of the El Borma-Gabes pipeline will be raised from 300 to
500 million cubic meters per year. The costs of TD 9 million will
be covered by Arab funds and STEG.[7]

As the gas is provided for the enterprises at capital costs, the
annual costs for 300 million cubic meters each will come to TD 0.9

million (extraction projected to last until 1986). However, there are
considerable differences between the possible production and the projected demand: while the utilizable production (the rest of the gas is
reinjected for the conservation of the deposits) will drop from 675
million cubic meters (1972) to 113 million cubic meters (1986) annually, the demand will increase from 150 million cubic meters to 500
million cubic meters, respectively.

In Gabes there is a thermoelectric production unit, powered by
El Borma gas since 1972, with a capacity of 60 megawatts, as well
as a gas turbine unit with a similar capacity. The latter (under
construction since February 1971) began production with one 15 mega-
watt turbine in September 1972. Two 22-megawatt turbines were
added in July 1973. In July 1975 construction began on an additional
gas turbine unit with a 60-megawatt capacity at Bouchemma, south
of Gabes. The total investment level was fixed at approximately TD
18 million (at the end of 1975). Thus, the cheap, long-lasting, and
comparatively ubiquitous energy supply of the Tunisian south is guaranteed. The gas production sold by El Borma already increased
from 0.7 million cubic meters in 1972 to 62.7 million cubic meters
in 1973.[8] Since April 15, 1973, ICM has also been a gas consumer
with 2,600 cubic meters per hour (22.8 million cubic meters annually).
This gas utilization had been planned for a longer period of time and
could only be accomplished after operation of the ICM power plant
had been changed from oil to gas. Unfortunately, no data could be obtained from the management concerning either the savings through this
alternative use of the gas or the share of the energy expenses in total
production costs. However, a study of the planned projects shows
clearly that the cheap energy will have a considerable influence on the
future industrialization policy of Gabes.

Projects realized or under construction are listed as follows:[9]

1. Second expansion stage of ICM (ICM 2) was inaugurated in
December 1974. At investments of TD 12 million, nearly 200 additional jobs were created. Construction began in 1972 and was financed
by French public and private capital. Additional production will be
130,000 to 150,000 tons of P_2O_5. In fact, production rose from (1974)
160,000 tons to (1975) 210,000 tons. In accordance with the Tunisian
state's new program to utilize the country's raw phosphates, a sodium
tripolyphosphate factory is under construction. The foundation stone
was laid in Ghannouche industrial park on March 22, 1975. The Al
Kimia Society will invest TD 4 million so that a capacity of 30,000
tons annually can be reached.[10] Total investment targets of ICM/Al
Kimia will be around $184 million, which should create a production
capacity of 900,000 tons of fertilizers annually.[11]

2. Expansion of the electric power plants. Gabes is the second
biggest electricity producer in Tunisia after Tunis. It will be con-

nected to the north of Tunisia by a power line carrying up to 225 megawatts. For the future, there are plans of a final production capacity
of 250 to 300 megawatts, unequaled in Tunisia outside its capital
(see also Chapter 13).

3. The foundation stone of the long-planned third Tunisian cement plant was laid in December 1974. The total investment costs
of the proposed Société des Ciments Portland of TD 29 million will be
predominantly financed by French public and private funds as well as
Libyan and Tunisian funds. The plant, which should have been completed by the end of 1975, will have a capacity of 2,000 tons per day
(approximately 700,000 tons of cement per year and 100,000 tons of
chalk). It will employ 276 persons.[12] The following reasons were
decisive in selecting Gabes as the location for the plant: favorable
transport location for the supply of the entire south, which consumes
approximately 25 percent of the cement production; existing raw materials (limestone, clay); location at the El Borma–Gabes gas pipeline; and a new harbor.[13] The plant is under construction 10 kilometers
from Gabes at the road to El Hamma.

4. The foundation stone of a fluorine production plant of the Industries Chimiques du Fluor was laid on April 5, 1974; adjoining
the ICM plant in Ghannouche. The necessary investments of $13.8
million (approximately TD 5.8 million) will be met by Tunisian and
French ($6 million) funds, public and private investors, as well as
from the World Bank ($0.6 million). Among the investors were also
the public Tunisian phosphate processing enterprises (SIAPE, ICM,
STEG, STEM). Production is supposed to begin in 1977, and an annual net revenue in foreign currency of $5 million is expected. The
yearly production capacity will reach 22,770 tons of aluminum fluo-
ride. Tunisian fluor-spar, imported aluminum, and sulfuric acid
from the ICM plant are being processed. Approximately 130 to 150
persons will be directly employed and an additional 90 to 110 indirectly, totaling 240 persons.[14] A further extension will consist of an
aluminum trifluoride production unit with a capacity of 10,000 tons
annually and is envisioned to cost $9 million.[15]

5. Gabes will also become the site of the country's second oil
refinery. Until now, the only refinery at Bizerte was just able to
match domestic consumption, but this demand will rise to 2.5 to 3
million tons of oil products in 1985; and the supply of the south is
rather costly. The capacity of the refinery will depend on whether
it will produce only for the domestic market (1.5 million tons per
year) or for export (6 to 7.2 million tons). Recent publications indicate that the construction of a first stage with a capacity of 3 million
tons should have begun in 1975. Investment costs would be TD 65
million. The plant eventually will be enlarged to 6 million tons.
Figures of an overall capacity of 7.5 million tons are also provided.[16]

Accordingly, the investment costs are supposed to range between TD 16.5 and 45 million (TD 22 million are provided by the Fourth Plan) and 200 to 800 jobs would be created. The project is still in the planning stage and will not be started before 1977. The location of Gabes can only be partly characterized as ideal. The new refinery would need a branch line from the TRAPSA pipeline (see Chapter 8), which would not have to be longer than 30 kilometers as the pipeline passes El Hamma. However, the capacity of the new harbor is already exhausted. An additional extension will be expensive because of the shallow water depth.

6. The construction of a monoamonium phosphate production plant is planned in Ghannouche on the basis of the phosphoric acid produced by ICM. A corresponding company, Ressources Tunisie S.A., with a capital of TD 1.2 million, was founded in October 1973 by U.S. enterprises and ICM. Construction was planned to start in 1975; further details are not yet known.[17]

7. Use of natural gas as the base for nitrogen production is also considered. In November 1975 a company for the production of phosphate and nitrogenous fertilizers, SEPA, was founded with capital participation of the United Arab Emirates. The total investments for production units in Gabes are estimated at TD 180 million.[18]

8. The brickyards of El Hamma (west of Gabes) will also be supplied by El Borma gas supplanting fuel oil. This plant of the Société tunisienne de matériaux de construction was developed out of traditional, private enterprises. It supplies the Southern Tunisian and Tripolitanian markets.

9. In early October 1973, a letter of intent was signed between the Tunisian government and German, Brazilian, and Japanese companies to construct a plant by 1978 for direct reduction of Brazilian iron-ore in the industrial zone of Gabes. Using natural gas, the plant should have been able to produce approximately 1 million tons of iron sponge. Investments of as much as $55 million (approximately TD 24 million) were foreseen. The transportation costs would have been reasonable, as the ore could have been transported as return freight of tankers that brought North African crude oil to Brazil. In addition, technological studies would have examined whether the plant could become the basis for a Tunisian steel production in the future.[19] (A small steel mill using traditional processes exists in El Fouladh near Bizerte). Actually, the project seems to have been canceled.

10. The project of a lime-processing plant is connected with raw material resources and will also be supplied by energy from the natural gas pipeline. At investment costs of TD 0.5 million, 120 jobs will be created; the annual production is supposed to reach 60,000 tons.

If all of these projects are realized, Gabes will without doubt develop into the most important industrial location, second only to Tunis itself. Investments of TD 71.4 to 117 million are planned just for the "safe" projects (1 through 6 above). They present a surprising amount of capital for the industrial sector compared to the total investments of TD 382.7 million projected in the Fourth Plan.[20]

Projects 1 through 5 above and the already-established ones will directly create at least 1,200 to 1,850 jobs, plus a number of indirectly employed contractors (who are estimated at 1,200 persons just for ICM). All of these relatively high-paid workers have a "multiplying effect" as associated with the oil industry (see Chapters 7 and 11).

This above-average importance of the projects is also clearly shown in the overall context of Tunisian development. Supplementing these new directly created jobs by the same number of indirect jobs (with all cautions), these approximately 3,600 jobs present a 6.1 percent share of the 58,600 jobs that are to be created in the Four-Year Plan; whereas the population of the Gabes Governorate makes up only 4.3 percent of the overall Tunisian population.

Associated Processes

All these projects have to be seen in connection with the development of the infrastructure. The El Borma gas pipeline, decisive for the energy supply, was already mentioned. Its full capacity (300 million cubic meters) is reached in 1976. If necessary, the capacity may be raised by additional pumps, which increase the pressure.[21]

The production and the transport of cheap energy was paralleled by the development of an entirely new harbor located at the chemical plant. Construction work started in 1969, the costs of approximately TD 10 million being financed to 85 percent with an Italian loan. Meanwhile, after the completion of the first stage, three ships of 15,000 tons and one of 50,000 tons can be handled at the same time. During the second stage, the harbor bottom is to be deepened for the handling of 100,000-ton ships. With this and the proposed airport south of Bouchemma (under construction in 1975-76, at costs of approximately TD 1.5 million), the infrastructure has improved significantly.[22]

Gabes will also be the main place of interest for the planned normal-gauge railroad from Sfax to Tripoli, Libya. It was projected in view of the quickly growing trade between Libya and Tunisia and the apparent economic integration between the Tunisian south and Tripolitania (Libya took 5.5 percent of Tunisian exports in 1975); it should have been completed by a conversion to standard gauge of the

Sfax-Tunis line. In the fall of 1974, a mixed Libyan-Tunisian railroad commission was founded. However, it does not seem very probable that the project (which was supposed to be financed by Libya),
will come into life soon because of the bad political relations between
the neighboring countries. More probable will be the realization of
a new railroad from Gafsa to Gabes (meter gauge as used in Tunisia
south of Tunis). It will be capable of transporting phosphates directly
(without the actual detour through Graiba, south of Sfax) from the
mines in the Gafsa-Metlaoui-Mdilla region to the processing plants.
Gabes being also the only public deep-sea harbor between Sfax and
Tripoli, its importance and locational favor will thus further increase.
It could also contribute to the supply of a part of Western Tripolitania,
which suffers from frequent congestion in the Tripoli harbor. [23]

The urban structure of Gabes will also change. As our studies
in El Borma showed (see Chapter 5), the Tunisian south has a relatively small potential of higher qualified workers required within the
modern industrial sectors. For the Gabes industrial location, this
means not only an exhaustion of the local work force potentials but
also the necessity of recruiting specialists, mainly from Northern
Tunisia and also, in a lesser degree, from abroad. [24] This again
would require the construction of living space, especially for the higher
paid specialists. Thus, in accordance with the 1973-76 Development
Plan, 636 dwellings, 62 "villas," and 77 apartments were completed
in 1973 and 1974 by national and local building enterprises. In addition, TD 6.4 million were invested in the construction of public buildings (schools, hospitals, administration buildings). Another 2,000
dwellings are to be constructed in Gabes during this planning period. [25]
All of these projects are based on the demand of the industry in Gabes.
Therefore they are closely related to the concept of the "growth-pole"
Gabes.

Various factors were and are decisive for the actual development
of Gabes as an industrial growth-pole center. Hydrocarbons as an
energy source and a future raw material become increasingly important. Thus, regional effects of the oil industry are relevant in the
Tunisian south: not in transport-unfavorable Medenine, nor in isolated Skhira, but rather in Gabes where several favorable factors come
together.

                                BIZERTE

So far, the only oil-processing enterprise in Tunisia is the refinery of the Société Tunisio-Italienne de Raffinage (STIR) in Bizerte.
Operations started on December 14, 1963. As with SITEP, branches
of the Italian ENI and the state of Tunisia each held 50 percent of the

company's capital, which was totally taken over by the Tunisian state in 1975. At first it processed imported oil, but since 1965 approximately 75 percent of the refined oil comes from El Borma. The processing capacity increased from 1 million tons (1965) to 1.25 million tons (1971). The production of (1971) 901,400 tons and (1975) 1,055,800 of oil derivatives is barely sufficient to cover the domestic market.[26] The plant employs 376 workers.[27] There are projects to raise the capacity (to 1.5 million tons annually by 1976) corresponding to the increasing demands of the domestic market. At a later date production may reach 3.25 or even 4 million tons.[28] In 1985 the internal demand in oil products will amount to at least 2.5 million tons, and accordingly a new refinery has been planned for quite some time. Tunis, Skhira, and especially Gabes were taken into consideration for the location. Gabes was finally decided upon. This location can be supplied easily from Libya and El Borma and its choice represents another step in the reduction of regional disparities in Tunisia (see previous pages).

The following factors were relevant to the choice of Gabes: the work force potential; the developing chemical industries, which can further utilize a part of the refined products; the existing material (dwellings, roads, railroad, harbor under construction) and educational infrastructures (schools); and the close distance to Libya as potential customers (although now supplied by the refinery in Zawiya, Libya). Without doubt Gabes is the best location in the Tunisian south for a concentration of industries based on energy. It will strengthen considerably the regional and locational positive effects of the industrial facilities already existing or under construction.

The refinery in Bizerte has not attracted further associated industries. Further installations of the petrochemical industries do not exist in Tunisia. In August 1968 an agreement was signed with Libya according to which Libya should concentrate on petroleum processing proper (ammonia and derivatives), while Tunisia would undertake the transformation of its phosphates to superphosphates, phosphoric acids, and artificial fertilizers.[29] Besides the existing superphosphate factories in Sfax, facilities of this kind are mainly projected for Gabes.

NOTES

1. According to Gabes 73—D'un centre agricole à un pôlé de developpement (Tunis: Direction de l'Aménagement du Territoire, 1973), p. 9, including comprehensive data especially of the urban planning of Gabes.

2. All figures from Les Villes en Tunisie (Tunis: Direction de l'Aménagement du Territoire, 1971), pp. 439-41, 450, and 459.

3. <u>Economic Yearbook of Tunisia 1966-67</u> (Tunis, 1968), p. 65 and 68.

4. All figures from "Gabes devient une capitale industrielle," <u>Europe France Outremer</u>, no. 513 (Paris), October 1972, pp. 36-38.

5. From <u>Marchés Tropicaux</u>, no. 1428 (March 23, 1973), p. 453; and no. 1437 (May 25, 1973), p. 1441.

6. Data from the Ministere du Plan, Tunis, 1973.

7. <u>Industries et Travaux d'Outremer</u>, April 1976, p. 265.

8. <u>L'Action</u> (Tunis), February 3, 1974.

9. Data from "Gabes devient une capitale industrielle," 1972; ICM lists; <u>IVe Plan de développement 1973-1976</u> (Tunis, 1973), pp. 263 and 268; "Le programme de développement de la STEG," <u>Marchés Tropicaux</u>, December 1975, p. 3676; <u>Industries et Travaux d'Outremer</u>, April 1976, pp. 290 and 297.

10. "Les projets industriels dans la région de Gabes," <u>Industries et Travaux d'Outremer</u>, no. 234 (May 1973), p. 453; and <u>Industries et Travaux d'Outremer</u>, April 1976, p. 290.

11. <u>Report in Investment Promotion Meeting and Technological Consultations for Chemical Industries in Developing Countries</u>, ID/WG. 197/5 of June 5, 1975 (Vienna: UNIDO, 1975).

12. <u>Marchés Tropicaux</u>, no. 1494 (June 28, 1974), p. 1951.

13. According to "Le ciment en Tunisie," <u>La Tunisie économique</u>, September-October 1973, pp. 12-15.

14. Among others, see <u>Marchés Tropicaux</u>, no. 1483 (April 12, 1974), p. 1006; International Finance Corporation, Press Release (Paris, July 3, 1974); and Ali Atia, "L'industrie extractive du spath fluor en Tunisie," <u>Conjoncture</u> (Tunis), no. 4 (July-August 1974), pp. 39-40.

15. <u>Report in Investment Promotion Meeting and Technological Consultations</u>.

16. <u>L'Action</u> (Tunis), March 7, 1975; <u>Petroleum</u> Economist, September 1975, p. 345.

17. <u>Marchés Tropicaux</u>, no. 1459 (October 26, 1973), p. 3214.

18. <u>Industries et Travaux d'Outremer</u>, April 1976, p. 290. In June 1976, the enterprise contracted the construction of a production complex with a capacity of 330,000 tons of phosphoric acid (P_3O_5) and 330,000 tons of di-ammonia fertilizer annually, at costs of TD 43 million. When operations start in 1978, 700 jobs will be created. At a latter stage, with 1,500 collaborators, production may more than double after 1980 and will be destined for exportation. See <u>Industries et Travaux d'Outremer</u>, no. 272 (July 1976), p. 541.

19. "Projet de création d'une usine d'éponges de fer à Gabès," <u>Marchés Tropicaux</u>, no. 1501 (August 16, 1974), p. 2356.

20. <u>IVe Plan</u>, p. 237.

21. Ibid., p. 248.

22. "Les objectifs du IVeme Plan dans le domaine aéroportuaire," Revue tunisienne de l'Equipement, no. 8 (April–June 1974).

23. See "Un chemin de fer Tunisie–Libye," Industries et Travaux d'Outremer, no. 248 (1974), p. 650.

24. See also Gabes 73, p. 106.

25. "Arrondissement de Gabes—Réalisations récentes et objectifs du Département pour la durée du Plan quadriennal 1973–76," Revue tunisienne de l'Equipement, no. 8 (April–June 1974), pp. 15–26.

26. Tunisia: The Development of the Petroleum Industry, E/CN. 14/EP/58, September 24, 1973 (New York: UN Economic and Social Council, 1973), p. 66; and Statistiques financières, no. 40 (May 1976).

27. According to lists in the Ministry of Planning, Tunis.

28. See IVe Plan, p. 251; and Industries et Travaux d'Outremer, April 1976, p. 286; Petroleum Economist, September 1975, p. 346.

29. Tunisia: The Development of the Petroleum Industry, pp. 67–68.

CHAPTER 10

REGIONAL EFFECTS OF
OIL-PROCESSING
INDUSTRIES IN ALGERIA

According to the striking difference in developmental concepts between Tunisia and Algeria (see Chapter 3), Algerian industrial development is more important than the activities shown in Tunisia.

ARZEW

Arzew, situated on the coast between Oran and Mostaganem, is an example for the systematic establishment of an industrial growth pole in Western Algeria. The starting point for the industrialization was the 24-inch natural gas pipeline from Hassi R'Mel (see Chapter 6) and the gas liquefaction plant based on this pipe of CAMEL (Compagnie Algerienne du Methane Liquide), under construction since September 1962. At that time the then president of the state had already mentioned Algeria's plans to develop Arzew as a center of petrochemical industries.[1]

Facilities of the Oil Industry

In the meantime, the following installations exist or are under construction:[2]

1. CAMEL, the world's first gas liquefaction plant, was built with English and French capital in 1962 (at a cost of DA 450 million) and started production in September 1964. Today, SONATRACH's share is 48.8 percent. The liquefaction capacity was increased from (1965) 1.5 billion cubic meters per year to (1968) 2.5 billion cubic meters of natural gas in accordance with the 15-year contract. The

entire production is delivered to France and Great Britain. Today, the formerly technologically leading installations are regarded as rather out of date; new facilities demand 30 percent lower investment. Therefore, it is possible that the plant may be shut down in 1979 after the expiration of the sales contracts and the amortization of investments. Currently, 308 workers are employed. Whereas 50 percent of the work force were foreigners in the beginning, today only 8 remain.

2. The SONATRACH factory for ammonia and nitrogenous fertilizer is also supplied by the natural gas pipeline. Construction began in 1966, the first production of artificial fertilizers was in 1969, of ammonia in 1970. The end products are (daily capacity) 1,000 tons of ammonia (NH_3), 500 tons of ammonia-nitrate (NH_4NO_3, for nitrogenous fertilizer), 400 tons of nitric acid, and 400 tons of uric acid ($CO[NH_2]_2$). Originally considered mainly for export, the entire production today (except for part of the uric acid production) is consumed domestically. The consumption of artificial fertilizers in Algerian agriculture, for example, increased from (1964) 6.5 kilograms per hectare to (1973) 37 kilograms per hectare and will probably reach 100 kilograms per hectare in 1980, necessitating additional imports. This was one of the reasons for SONATRACH's order in April 1974 to construct two additional units with a daily capacity of 400 tons of nitric acid. The plant employes 780 persons.

3. In addition to the gas pipeline, the construction of an 801-kilometer oil pipeline from Hassi-Messaoud was started in 1964. With a diameter of 28 inches and with six pumping stations it has an annual capacity of 20 million tons. Since the foreign companies operating in Algeria were not willing to run the risk of the investment, this oil pipeline was the first major project in Algeria to be exclusively contracted and financed by SONATRACH. The shipment from this pipeline (1970: 19.96 million tons; 1971: 15.08 million tons; 1973: 19.3 million tons) started in February 1966. Arzew became quantitatively the most important export harbor of Algeria, followed by Skikda. The pipeline provides work for approximately 650 persons of whom approximately 400 work at the Arzew terminal.

4. Based on these installations, a Japanese company was contracted to construct the country's third refinery in Arzew, which supplies Western Algeria.[3] Inaugurated in June 1973, the refinery has an input capacity of 3.5 million tons of crude oil annually and produces 30,000 tons of ethane and methane; 110,000 tons of propane and butane; 350,000 tons of regular and premium gasoline; 400,000 tons of naphtha; 150,000 tons of kerosene; 150,000 tons of gas oil; 950,000 tons of heavy fuel oil; 65,000 tons of bitumen; and 53,000 tons of lubricating grease. Of this production, gasoline export approaches 70 percent, gas oil 50 percent, and naphtha 100 percent. Originally, the construc-

tion costs ran up to $70 million. A work force of 703 persons is
needed to operate the plant; during the construction period 3,500
workers were employed, encompassing more than 5 million working
hours. An increase in the processing capacity to 5 million tons annually is possible. The first step taken in this regard was a contract
with a Japanese enterprise, in September 1975, to raise the production of bitumen from an original 65,000 to 140,000 tons per year, at
investment costs of DA 38 million. [4]

5. The plant for the separation of oil gas (LPG, liquid petroleum
gas) was to start operation along with the refinery and after the completion of a 16-inch condensate and LPG pipe from Hassi Messaoud.
This, however, did not happen before the end of 1973. The initial
capacity was 850,000 tons of LPG and 1 million tons of condensate.
The condensate capacity may later be raised to 3 million tons. The
LPG (approximately 1 million tons per year), which until that time
had been burned in Hassi Messaoud (see Chapter 6) is being disasso-
ciated in Arzew into its components of propane (1,900 tons per day)
and butane (1,100 tons). Two-thirds of these products are exported.
The plant employs 400 persons.

6. The production complex for methanol and synthetic resins
is near completion. The construction of the methanol plant began in
1972, and its capacity is approximately 300 to 400 tons daily or
130,000 tons annually. The plant itself uses 20,000 tons, the rest
being destined for domestic use (30,000 tons) and for export (80,000
tons). The synthetic resins unit annually will produce 20,000 tons
of formaldehyde (36 percent); 2,500 tons of phenolic resins in powder; 1,800 tons of liquid phenolic resins (85 percent concentrate);
1,600 tons of liquid phenolic resins (45 percent concentrate); 5,500
tons of urea resins in powder; 5,500 tons of urea glues; 500 tons of
melaminic resins; and hardeners for urea glues. During construction, 1,000 persons are employed; when completed, it will be operated
by 400 employees, including 30 managers and engineers, and 90 technicians or foremen.

7. At the same time the refinery was inaugurated, the foundation stone of the first section of the new gas liquefaction plants (two
units) was laid in Arzew at total investment costs of DA 3.4 billion.
This will later be extended to four units that are to treat 15.5 billion
cubic meters of gas. [5] Although not all of the extension projects have
yet led to firm contracts with construction measures, Algerian officials are confident of the completion of the projects. A second 10-
inch natural gas pipeline from Hassi Messaoud is under construction.
It was also decided to raise the capacity from 6.5 billion cubic meters
per year to approximately 13.5 billion cubic meters per year by the
installation of five pumping stations. After completion of the supple-
mentary utilities, a total of 27 billion cubic meters of natural gas per

year are planned to be liquified. The costs for all utilities, including
the natural gas pipes, are estimated at DA 7 billion. The export of
liquefied gas to the United States should start in 1976 (see Chapter
12).

8. The existing harbor is not even able to cope with the present
activities, not to mention the future. Therefore it was decided to
construct a new port at Bethioua, ten kilometers east of Arzew, to be
called Arzew el Djedid (New Arzew). A projected additional gas
liquefaction plant is also to be constructed there.[6] The new harbor
will be three times larger than the harbor of Skikda (see below) and
will be able to accommodate simultaneously six 125,000-200,000 cubic
meters capacity liquefied natural gas (LNG) tankers, three oil tankers
of 100,000, 150,000, and 200,000 ton capacities, respectively, and
one ammonia transport ship. The total shipment of LNG is supposed
to reach 40 billion cubic meters annually. The costs of the entire
project are estimated at $300 million (DA 1 billion).

9. As the demand for water in the industrial plants will increase,
the construction of a seawater desalination plant has been commissioned. It will produce 80,000 cubic meters of water daily and thus
will be able to provide fresh water for the entire town of Arzew and
most of the big production units.

Effects on the Labor Market

The utilities that have been completed already offer more than
3,000 industrial jobs with comparatively high incomes. Currently
at least 4,000 additional construction workers are employed. After
the completion of the gas liquefaction plant, the former figure will
rise to 3,700 persons. According to a different source, in the autumn
of 1972, 8,980 workers were employed at the industrial plants proper
and at the construction sites. Approximately 40 percent of the employees at the SONATRACH plants live in Oran, another 10 percent in
Mostaganem, and the rest found accommodations in Arzew and its
vicinity. A comparison of these employees to the population of Arzew
(approximately 20,000 in 1974, see below) clearly shows that the
town's work force potential, as well as that of its surroundings, is
more than exhausted, as depicted in the next section.

Associated Processes

To a much greater extent than the other harbor towns (such as
Skikda, Bejaia, or Gabes), Arzew can hardly cope with the demands
emanating from an ever-growing population.[7] Living, before indepen-

dence, from the harbor and fishing activities as well as from a small
sulfur refinery (offering 100 jobs), the population grew from (1959)
4,000 to (1963) 7,000. It then doubled (by 1972) to approximately
15,000, may have reached 20,000 in 1974, and will grow to at least
45,000 by 1980. The traditional core of the city, so far maintained,
is completely overburdened. New residential quarters are now developing at the periphery of the old center. Although the demand for new
lodgings is estimated to be at least 10,000 by 1980, not more than
1,100 are under construction or planned. The infrastructural facilities necessary for an industrial city of this size are also lacking.
For instance, the nearest hospitals are in Oran and Mostaganem. So
the infrastructural development is lagging behind industrial development by approximately three to four years. For instance, no structures are available for the necessary enlargement of the post office,
schools, administration, customs office, courthouse, and similar
institutions.

Because of this, many of the workers who became recently employed in Arzew live in Oran and Mostaganem. They commute daily
90 kilometers (round trip). The short distance to Oran where reserves
of housing space doubtlessly still exist made Oran a "dormitory" for
the industrial workers of Arzew. Subsequently, the secondary effects
of industrialization on Arzew are absolutely lesser than, for example,
in Skikda. Relatively, however, they were certainly important with
regard to the smaller population that had been living there before. [8]

Even the necessary transport facilities for the traffic flows,
caused by the commuters and industrial in- and outputs, have to be
created. The Oran-Arzew road is being expanded to four lanes.
The Oran-Arzew highway with a later extention to Algiers will be included in the next Four-Year Plan. A new railroad is to connect the
two cities with a capacity of 5,000 passengers per hour. It will link
Arzew with the harbor of Oran and is to be extended to Mostaganem.
A big marshaling yard for freight cars is projected east of Arzew at
Marsa El Hadjadj. Only a close link to Oran within a common development axis can solve the future problems of Arzew, as only Oran
(actual population of 550,000, predicted to be 851,000 in 1985) is
able to provide facilities for an industrial labor force with higher
demands. In the framework of the concept of a "linear urbanization"
along the Oran-Arzew axis, it has to be assumed that—on a long-term
basis—the two cities will grow together. [9]

SKIKDA

Skikda is another example of a radical transformation of a city
through the effects of the oil industry. Until 1968 Skikda was a sleepy

rural town without dynamic development. The then approximately
66,000 inhabitants mainly lived from the food-processing industries
and harbor activities. The latter had been stagnant since 1957. In
1967, for instance, the harbor's turnover consisted of 88,604 tons
of export items (mainly scrap metal, dates, and vegetables) and
273,502 tons of import items (cereals, petroleum products, and pipes
for oil pipelines).[10]

In the hinterland, agriculture remained practically the sole
source of income. The only nonagricultural enterprises were the
marble quarries of Filfila, which have also been stagnant since independence. Not before their modernization in 1971 did they start expanding again. Today, 158 persons are employed in the quarries,
and another 140 persons in the new stone processing plants.[11]

Facilities of the Oil Industry[12]

In March 1967 the Algerian government decided to construct a
natural gas pipeline from Hassi R'Mel to Skikda, and an oil pipeline
(in parts parallel to the former) from Haoudh El Hamra (Hassi Messaoud). At a diameter of 40 inches and a length of 573 kilometers,
the gas pipeline has a basic capacity (without pumping stations) of
6 billion cubic meters annually. In 1975 a second developmental
stage (two pumping stations added) increased the capacity to 9 billion
cubic meters. A third stage (three pumping stations) will enable the
pipeline to transport 12 billion cubic meters of natural gas.

An additional parallel 40-inch gas pipeline is projected and will
be necessary when all of the liquefaction units start operating. The
oil pipeline has a length of 637 kilometers, a diameter of 34 inches,
and a basic capacity of 12 million tons per year (the Mesdar-Haoudh
el Hamra section is 107 kilometers, 26 inches). This capacity
might be increased to 18 to 24 million tons in 1976, and later to 30
million tons. Both pipelines were inaugurated in May 1972. So far,
the oil pipeline transports oil only to be exported, shipment still taking place in the oil harbor.

The construction work on the pipelines was paralleled by the
decision to develop a petrochemical industry.

1. In the beginning of February 1971, SOMALGAZ, the holding
company for a gas liquefaction plant, was founded with participation
of French companies; at the end of 1971 it was completely taken over
by SONATRACH.[13] The liquid gas, produced in the first three stages
of the plant, is to be exported to France via Fos-sur-Mer (near Mar-
seille). In 1969, the French company TECHNIP started construction
on the first three units of the plant (liquefaction capacity of 4.5 billion

cubic meters of natural gas). Production began in December 1972.
The fourth unit, built by Pritchard Rhodes, was completed in 1975.
Two additional units were ordered to be built by Pritchard Rhodes in
July 1973. Altogether, the six units will generate 90 billion thermal
units per year, requiring approximately 9 billion cubic meters of
natural gas.[14] The plant should be completed at the beginning of 1977.
Aside from the liquefied gas, it will produce approximately 270,000
tons of ethane (a raw material for chemical industries) and 440,000
tons of butane and propane (to be partly bottled). In the first developmental stage (four complexes) 502 persons, including 30 engineers
and 90 technicians and foremen, are needed, another 160 in the second.
In addition, up to 3,000 persons are employed on the construction
sites. The produced LNG, which is stored at -162°C, is to be shipped
in special tankers to Europe and the United States (see Chapter 12).
The overall investment costs (first stage) amounted to DA 900 million.
Due to technical problems, the production target of 4.5 billion cubic
meters could not be reached in 1974. But it seems that the initial de-
fects have now been eliminated.

    2. The terminals of the oil and natural gas pipelines have al-
ready been completed.[15] Until the inauguration of the new harbor, the
oil is shipped from the old harbor in Skikda (which, because of security
reasons, has been closed to the public). Approximately 485 workers
(including 55 executives and 120 technicians and foremen) are required
to operate the oil pipeline, of which 80 work in Haoudh el Hamra and
40 at the Biskra pumping station. Investments for the installations by
SONATRACH totaled DA 781 million.

    3. In order to supply the industrial zone and the town of Skikda
and its vicinity with electrical power, the national electricity company
SONELGAZ constructed a natural gas power plant with a capacity of
140 megawatts. It started operation in 1975 and was connected to the
national electric network. Since 1975 a second 190-megawatt unit
has been under construction.[16] An enlargement to 800 megawatts
is planned for 1980. Currently, the main plant employs approximately
150 persons.

    4. Since 1972, a production plant of raw material to be used
in the plastics industries has been under construction on an area of
48 hectares. After completion it will annually produce 120,000 tons
of ethylene using the ethane produced by the gas liquefaction plant;
48,000 tons of polyethylene; 40,000 tons of sodium chlorate; 40,000
tons of monovinyl chloride; 35,000 tons of polyvinyl chloride (PVC);
1,000 cubic meters per hour of fresh water (produced by the water
desalination unit). The plant (projected investment costs of DA 8
billion) should start production in 1977 and employ approximately
1,000 persons.[17] It will provide the raw material for the plastics
production center in Setif (see below), which still depends on imports.

    5. In April 1974, an Italian company was contracted to construct
a refinery with a capacity of 15 million tons per year[18] (originally,
a capacity of 7.5 million had been projected). The plant is to be completed in early 1978 at costs of $300 million, and will annually produce:
100,000 tons of propane; 390,000 tons of butane; 3,500,000 tons of
naphtha; 720,000 tons of gasoline; 46,000 tons of kerosene; 5,000,000
tons of gas oil; 4,680,000 tons of fuel oil; 25,000 tons of bitumen;
95,000 tons of benzol and toluene; 185,000 tons of xylene and para-
xylene. The fuels will be exported; the aromatic products and a part
of the naphtha will supply the petrochemical complexes in Skikda and
Setif. With the above refinery as well as the two other already existing ones (Arzew and Algiers), Algeria will be able to process half of
its domestic oil. The Skikda plant will provide jobs for approximately
1,000 workers.
    6. Among the smaller projects are a facility to store and process
LPG and a propane and butane bottling plant already in operation.
Capacities are at 300,000 and 235,000 tons, respectively, and 250
persons are employed.
    7. A supply and security base (approximately 100 employees)
is part of the entire complex. Further lots are designated for additional processing enterprises.
    8. In November 1975, tenders for an ammonia production unit
in Skikda were published. Its desired daily capacity of 1,000 tons is
to be consumed in Algeria. However, Annaba is also considered a
possible location for the unit.[19]
    9. The construction of a new harbor was indispensable for the
industrial plants and was completed in 1974 at investment costs
of DA 440 million. The harbor has three berths for 50,000- to
100,000-ton tankers, two berths for 125,000-cubic meter LNG tankers, and one berth for ammonia tankers.

    The entire industrial zone (1,924 hectares) was expropriated by
decree on June 18, 1973.[20] This included the previous airfield (180
hectares), roads, streets, and rivers (91 hectares), and 1,578 hectares of private property, mostly formerly owned by Europeans. A
total area of 3,000 hectares is planned to be developed for all industrial projects.

                       Effects on the Labor Market

    The above-mentioned plants directly provide more than 1,000
highly paid jobs. This number will increase to approximately 3,400
by 1978 when the contracted projects will be completed. According
to another source, approximately 8,000 permanent jobs are to be

created by 1980. This would mean that 27 percent of the then probably 150,000 inhabitants will live on incomes from SONATRACH and associated enterprises, assuming a work force of 30,000 persons (activity rate of 20 percent). In the long-term range, according to the will of the Algerian planners, Skikda shall become, by 1985, "one of the world's biggest centers for petrochemical and natural gas industries." Its population will then reach 200,000 people, and the then 10,000 industrial jobs will generate another 30,000 qualified job offerings in related sectors of the regional economy.[21] In fact, in mid-1973 SONATRACH directly employed a work force of 2,500 workers. Another 1,500 to 2,000 persons were employed by contracting firms in the industrial zone. This causes a basic change in the income structure of the whole region.

Interviews revealed that three (15 percent) of the interviewed had been unemployed before they joined SONATRACH; the others had jobs but their salaries were generally one-third lower.[22] So far, the labor could be recruited within the region, although some of the specialists come from the big urban centers. Previously, some of the interviewed had worked in agriculture and others in trade professions. The transformation affects the entire region. Jobs are created for former under- or unemployed people as well as for the increasing number of better-educated and trained Algerian specialists and engineers.

The interviewed (average age of 31.3 years) earned an average of DA 1,180 per month. They came predominantly from the city of Skikda (eight persons or 42 percent), from villages in the vicinity (six persons or 31 percent) and the Kabylia region. The problems of unemployment in this traditional emigration region (four or 21 percent of the interviewed had previously worked in France) were as urgent as in the Sahara. Even today, each interviewed worker has to support an average of 6.7 persons. The effect of these industrial jobs is intensified by the high income level of SONATRACH workers. They compare, for instance, with the wages of dockers in Skikda harbor, which, reaching DA 630 per month, already exceed the Algerian average.[23]

Except for the people directly employed by SONATRACH, a large number of private and public enterprises work as SONATRACH contractors in the industrial zone. These include civil engineering (600 workers), steel construction (150 workers), painting (300 workers), and hydraulic construction (500 workers) with a total of at least 1,550 employees. This amount might even reach 2,000 persons temporarily. For instance, in 1972 SONATRACH spent DA 8 million for services rendered by contractors.

Associated Processes

The processes associated with the development of an industrial center are manifold. All of eastern Algeria has benefited from the introduction of natural gas. Energy is provided by means of 1,000 kilometers of branch pipes for enterprises and private households in the towns of Biskra, Batna, Ain M'lila, Constantine, Azzaba, Guelma, El Khroubs, Souk Ahras, Annaba, El Kala, El Milia, Jijel, Setif, Bordj Bou Arreridj, and Bejaia.

However, numerous problems also emerge for the city of Skikda and its immediate vicinity.[24] Draining the labor force from agriculture causes food prices to increase above the general Algerian level. Correspondingly, the supply of fresh food is sometimes inadequate.

The fishing fleet had to leave its assigned part of the harbor for security reasons and subsequently was laid idle. As a result, the fish-processing factories with a work force of several hundred persons had to close down too. The housing situation was also unfavorably affected. As the housing capacity of the city is only sufficient for approximately 60,000 inhabitants (in 1974 the population was estimated at 130,000 inhabitants), 35,000-40,000 had to live in provisional shacks (gourbis). This marginal population, attracted by the mushrooming industries, is neither physically nor psychologically integrated into the city. They generally exist on temporary employment as unskilled laborers within contracting enterprises.

The industrial growth pole of Skikda, whose purpose is to exert a dynamic impact upon the other sectors of the regional economy, may have developed too quickly. The petrochemical industry is a good example for an industrializing industry, as it is supposed to create for each of its own jobs three to four additional jobs in the construction business, the service sector, and associated industries. However, our regional survey depicts the generally small elasticity within the quality of work force of the labor market in regions outside the big cities.

The resulting heavy migration into Skikda has aggravated the housing problem the most. Currently, approximately 5,000 families are estimated to be affected by the housing shortage. The construction of 1,800 housing units is projected, and in 1976 an emergency program for 1,000 units was started. However, no more than 335 had been built by early 1974, although 1,130 dwellings were allegedly under construction in 1973. The main problem is the lack of capable enterprises and labor, all of whom are generally needed in the industrial zone. In contrast to Arzew, there are no bigger cities near Skikda (80 kilometers to Constantine, 100 kilometers to Annaba) that would be able to accommodate some of the immigrants. This quick population growth even caused certain difficulties for the private and

public sector in the supply of goods. The housing shortage is so se-
rious that SONATRACH now only employs workers who can prove that
they have accommodations in Skikda. So far, the development of the
urban infrastructure has not been able to keep up with increased demand.

SETIF

The interplay between the natural resources of the oil sector,
the capital that is at the disposal of the country, and the labor potential (especially in the so-far scarcely industrialized highlands and
steppe regions) may be arranged in order to bring about enormous
projects. These projects are at the same time being integrated into
the remaining sectors of the economy. Thus, one of the aims of Algerian economic policies—"the fight against economic disparities"—
can be taken on.

Facilities and Projects of the Plastics Industry

Since October 1971, the following SONATRACH projects were
planned or became operational. Raw material will be provided from
the Skikda plants (see above), and the investment target lay around
DA 800 million in 1974:[25]

A plastic-item factory (low pressure polyethylene) with a capacity
of 30 million plastic bags annually (8,000 tons), 3,800 tons of plastic
films and tarpaulins, and 2,300 tons of plastic milk containers. In
the summer of 1971 a German enterprise was contracted to construct
the factory (requiring investments of DA 35.4 million). Production
started in January 1973 with approximately 210 employees.
A plastic pipe factory (polyvinyl chloride, PVC)—administered
by the Societe Nationale des Materiaux de Construction, SNMC—annually produces 2,400 tons of pipes with diameters ranging from 20
to 200 millimeters. In 1973 the production was faced with a domestic
consumption demand of 4,700 tons. In 1971 an Austrian company had
been contracted to construct the factory, and it was completed in December 1973. Production also began that year employing 150 workers.
(Total investments for the project amounted to DA 22.3 million.)
A floor-covering factory (polyvinyl chlorite) with an annual capacity of 9,000 tons (1.6 million square meters). The construction
contract was commissioned in May 1972, and production began in 1974,
creating 180 jobs.

TABLE 10.1

Projects in the Plastics Industry in Setif

Production Sector	Capacity in Tons per Year	Jobs	Projected Start of Production
PVC foils and tiles	3,000	150	July 1976
Buckets	2,200	130	May 1975
Profiled pieces	2,500	150	Jan. 1976
PVC pipe extension (2nd)	5,400	80	Jan. 1976
Household appliances	2,000	250	July 1977
Package materials	2,000	300	Jan. 1977
Plastic, paper, cardboard	4,200	160	Feb. 1977
Pots and vessels	3,000	250	July 1977
Hygenic materials	2,000	180	July 1977
Children's articles	2,500	160	Jan. 1978
Plastic furniture	2,600	260	Sept. 1978
Parts for industry	3,500	140	Oct. 1977
Toys	1,000	120	July 1977

Source: "Centre plastiques de Setif," Europe Outremer, no. 533/534 (1974), p. 34.

A polyethylene mesh factory (the product being mostly used for agricultural purposes) with a capacity of 4,000 tons annually, employing 210 workers.

The above-mentioned factories are already completed or under construction. They represent only the first stage in the development of Setif as the center of the Algerian plastics processing industries. The projects listed in Table 10.1 are planned for the future. Altogether, these projects will create 2,330 jobs, so that the plastics processing sector alone should employ 3,080 persons in 1978.

The building-material sector developing in Setif is not connected to the petrochemical industry. It contains a plant for the fabrication of reinforced concrete parts (30,000 tons per year) and flagstones, the SONELEC dry battery and accumulator plant with 500 employees (which may also produce electronic instruments at a later stage), and a plant probably employing 890 persons in 1976 for sanitary mountings in neighboring Ain Kebira. The second biggest project after the plastics industry will be the SONACOME plant for industrial vehicles expected to offer 5,000 qualified jobs. Several modernized or developing food-processing enterprises supplement these projects.

Effects on the Labor Market

The enterprises completed in 1974 have already created 6,000 new permanent jobs. It is estimated that in 1976 30,000 persons will be directly or indirectly employed in the industrial zone. Considering average net wages of DA 975 per month, additional incomes of DA 351,000 will be available in the region as a result of the two already existing plastics processing plants.

In contrast to the harbor towns of Arzew and Skikda the labor market in Setif seems to be at least quantitatively more than sufficient.[26] Of the city's 1972 population of 130,000 inhabitants, 20,000 (one-sixth) were unemployed. In the surrounding Dairates (counties), the situation is even more unfavorable. So far, the only solution has been emigration. In the small Daira Sidi Aich, for instance (population 6,900), 2,000 men emigrated to foreign countries. According to the 1966 census, the Wilaya Setif, after the Tizi Ouzou Wilaya, had the second highest number of men working in foreign countries (70,504 persons, which is 10.96 percent of the male and 5.71 percent of the entire population).[27]

The above-mentioned projects, in contrast to the plants of the petrochemical industries on the coast, are thus labor rather than capital intensive. Above all, they are based on the labor potential, which will further increase corresponding to the slowdown in labor emigration as desired by the government.

Associated Processes

The choice of Setif as the center for plastics-processing industries indicates the possibilities and directives of Algeria's national economy plans. Processing industries are independent of immobile locational factors as their products can be distributed easily. Thus, the development of processing industries on the high plateau and in the steppe regions helped to solve their inherent problems (high population pressure, unfavorable interregional traffic facilities, and lack of natural resources). According to an Algerian report from 1972, investments of DA 630 million were necessary for a polymers fabrication plant (basis for the plastics-processing industry), which would create 750 jobs (investment costs of DA 840,000 per job). An additional DA 510 million to be invested in the further processing of these polymers would, however, create 13,000 jobs (requiring an investment of DA 39,230 per job). The annual turnover will reach DA 143 million and DA 750 million, respectively. We see that the produced surplus value of the processing of raw material is much higher. So, the existing manpower potential was the starting point for the decision to make Setif a part of the eastern growth pole, closely interlinked with Skikda.

Effects of industrial activities are already visible. From 1969 to 1974 the total number of industrial workers (including construction) in the Wilaya Setif (which also included the Daira Skikda) increased from (1969) 8,106 to (1970) 8,622, to (1971) 12,197, to (1974) 14,903 workers. This is an increase of 83.8 percent (Algerian average: 62.1 percent).[28] However, corresponding to the temporary labor demand for construction sites, the number of construction workers increased from (1969) 1,899 to (1974) 6,762 persons.

Setif, too, shows the picture of a city in transition. New schools, hospitals, and dwellings are still able to accommodate the increasing number of industrial workers. But their demands are certainly lower than those of the workers in the basic petrochemical industries in Arzew and Skikda.

NOTES

1. See A. Benbella, "La politique pétrolière algérienne," Revue algérienne des Sciences juridiques économiques et politiques, no. 4 (1964), pp. 83-103.

2. Data given by the individual managements and listed literature; see also the survey in "Première carte industrielle de l'Algérie," Industries et Travaux d'Outremer, no. 247 (1974), pp. 546-62; and different SONATRACH information leaflets. For LNG see also Chapter 12.

3. "Zone industrielle d'Arzew—le triomphe de la volonté," Moudjahid, June 16, 1973; Moudjahid, June 18, 1973; "La raffinerie d'Arzew," Industries et Travaux d'Outremer, no. 234 (1973), pp. 451-52; "Le Président Boumedienne a inauguré une importante raffinerie à Arzew," Le Monde, June 19, 1973.

4. Marches tropicaux, September 26, 1975, p. 2684.

5. "Précisions sur la réalisation de la phase finale du gazoduc Hassi R'Mel-Arzew," Industries et Travaux d'Outremer, no. 237 (1973), p. 714; Marchés Tropicaux, no. 1505, September 13, 1974, p. 2584; most recent data from Petroleum Economist, March 1976, p. 113.

6. "La construction du nouveau port de Béthioua se prépare," Marchés Tropicaux, no. 1461, November 9, 1973, p. 3251; "Projet de'aménagement d'un nouveau terminal pétrolier et gazier à Béthioua," Petrolé et Gaz arabes, no. 123 (May 1, 1974); Moudjahid, June 18, 1975; Industries et Travaux d'Outremer, no. 272 (July 1976), p. 540.

7. For the following, see the already-mentioned literature and "Savez-vous comment on apelle Oran aujourdhui," Moudjahid, October

14, 1972; "Un futur centre agro-industriel et universitaire," <u>Moud-jahid</u>, July 14, 1973; "Arzew face à son avenir," <u>Algérie et Développement</u>, no. 16 (1973), pp. 12-15; "Ce que sera Oran en 1985," <u>Moudjahid</u>, March 26, 1974; "Complexe industriel d'Arzew," <u>Moudjahid</u>, January 15, 1975.

8. In contrast to the opinion of Dj. Sari, "La restructuration des centres urbains en Algérie," <u>Annuaire de l'Afrique du Nord 1972</u> (Paris, 1973), pp. 55-75.

9. J. Franchet, "La formulation de l'espace algérien," <u>Annuaire de l'Afrique du Nord 1972</u> (Paris, 1973), pp. 39-53.

10. Mme. Grangaud, "Les activités portuaires comparées de Skikda et Annaba," <u>Annales algériennes de Géographie</u>, no. 6 (1968), pp. 36-39.

11. Y. El Kenz, "De grandes perspectives pour l'usine de marbre de 'Fil-Fila," <u>Moudjahid</u>, October 23, 1974.

12. According to information given by the individual plant managements, by the Coordinateur de la Zone industrielle, from quoted literature, and SONATRACH information leaflets; see also the survey in "Première carte industrielle de l'Algérie," <u>Industries et Travaux d'Outremer</u>, no. 247 (1974), pp. 546-62.

13. These and the following data are from SONATRACH, ed., <u>Usine de liquéfaction de gaz natural de Skikda</u> (Algiers, 1972); "L'usine de liquéfaction de Skikda," <u>El Djeich</u> (Algiers), no. 142 (March 1975), pp. 21-24.

14. See, among others, "Accord SONATRACH-Pritchard Rhodes," Moudjahid, July 7, 1973; as well as "Pritchard Rhodes construira deux autres unités de liquéfaction à Skikda," <u>Industries et Travaux d'Outremer</u>, no. 237 (1973), pp. 713-14.

15. SONATRACH, ed., <u>Oléoduc Mesdar—Haoudh el Hamra—Skikda</u> (Algiers, 1972), p. 5; and <u>Moudjahid</u>, June 18, 1975.

16. <u>Industries et Travaux d'Outremer</u>, no. 249 (1974), p. 718.

17. "Skikda—Le complexe pétrochimique prend de l'extension!" <u>Moudjahid</u>, March 7, 1974.

18. See "Snam Progetti et Saipem chargées de la construction d'une raffinerie de 300,000 b/j à Skikda," <u>Pétrole et Gaz arabes</u>, no. 124 (Beirut) (1974), May 16, 1974, pp. 9-10; also "Contrat signé entre la Sonatrach . . . " <u>Marchés Tropicaux</u>, no. 1487 (May 10, 1974), p. 1248.

19. "L'industrialisation de l'Afrique—les nouvelles usines et les grands projets en 1975," <u>Industries et Travaux d'Outremer</u>, no. 269 (1976), p. 288.

20. Published in <u>Moudjahid</u>, June 18, 1973.

21. For the following, see R. Afredj, "Skikda—les retombées du boom industriel," <u>Moudjahid</u>, August 3, 4, 6, 1974; Y. El Kenz, "Zone industrielle de Skikda—40,000 emplois séront créés d'ici à 1985," <u>Moudjahid</u>, June 29, 1976.

22. Random spot check of 19 SONATRACH workers (two of whom were women), June 1973.

23. "L'intégration des dockers a la SONAMA," Moudjahid, April 30, 1974.

24. For this and the following thoughts, see Afredj, "Skikda"; "Skikda—une nouvelle cité pour 476 familles," Moudjahid, August 1, 1973; "Skikda face aux bouleversements économiques," Moudjahid, February 15, 1974; P. Balta, "L'Est algérien à l'heure du 'boom' industriel," Le Monde, April 19 and 20, 1974; "Les approvisionnements à Skikda," RUZIKAD (SONATRACH), no. 5 (March 1976), pp. 13-14; and El Kenz, "Zone industrielle de Skikda."

25. Data from SONATRACH, ed., Une nouvelle industrie de pointe en Algérie—le plastique (Algiers, 1972); "30,000 emplois nouveaux d'ici à 1975," Moudjahid, May 14, 1974; in general also "Première carte industrielle de l'Algérie," Industries et Travaux d'Outremer, no. 247 (1974), pp. 546-62; R. Benattig, "La SONATRACH à la veille du 2ème plan quadriennal," Algérie Actualités, no. 438 (March 10, 1974); and SONATRACH information leaflets; for the development of the Algerian plastics processing industry see, for instance, "L'industrie algérienne des matières plastiques veut rapidement égaler celle des pays les plus avancés," Marchés tropicaux, no. 1401 (September 15, 1972), pp. 2657-58; and M. Souidi, The Development of the Plastics Industry in Algeria, ID/WG.189.3, November 12, 1974 (Vienna: UNIDO, 1974).

26. For the following, see also R. Afredj, "La Wilaya de Sétif mobilisée pour son essor économique," Moudjahid, November 1, 1972; and A. Tiouri, "Sétif: Agriculture et industrialisation," Moudjahid, July 8, 1976.

27. Annuaire statistique de l'Algérie, 1970 (Algiers, 1971), p. 27.

28. Annuaire statistique de l'Algérie, 1974 (Algiers, 1975), p. 38; and "Les resultats (partiels) de l'enquête emploi et salaries de 1974" (Algiers: Secretariat d'Etat au Plan, 1975).

SYNOPSIS: OIL
PROCESSING AND
REGIONAL DEVELOPMENT

The contents of the previous chapters showed that the spatial effects of the oil industry are manifested only in specific locations and under specific conditions. The economic constraints inherent in the oil industry (resulting from the conditioning factors of geology and technology as well as from the mechanisms of the domestic and international markets) permit a certain systematization, but not a quantification, of these effects.

In Tunisia, regional effects have so far been relatively small, as a result of the probable absence of development strategies that regard hydrocarbons as a factor to be integrated into economic planning (see Chapter 2). Therefore, the following statements will be mainly restricted to Algeria.

EFFECTS ON SPATIAL STRUCTURES

As shown in Chapter 1, positive regional effects of mining industries can be realized only when such enterprises in the developing countries are no longer looked upon as merely suppliers for the industrialized countries. This is not in contradiction to the stated long-term "fugitive character" of the mining industry, as the fugitiveness is only relevant at the prospecting and production level. Processing exerts a long-term impact, and as oil belongs to the irreplaceable and decreasing natural resources, special importance has to be attached to development and differentiation in higher levels of the processing industries. The reason those "primary" effects of the oil industry have hardly been acknowledged is that they have only recently become relevant, that is, after the producing countries gained full control of their raw materials.

In Tunisia and Algeria the concept of a domestic oil-processing industry did not come to life before the initiation of their own development strategies. The Tunisian efforts began very slowly with the Perspectives decennales in 1962. In the same year newly independent Algeria started its development program. At first only the harbor towns benefited from these concepts as they were the locations of foreign trade-oriented production. Today almost all of the Algerian harbors are locations of an industrial development. However, the regional planner has to ask whether the exclusive development of industries in harbor towns does not aggravate the risks of accelerating the depopulation and relative impoverishment of the interior of the country.[1]

Historically, the precolonial centers of population and activity had been off the coast, and during the French colonization North African seaports were mainly big immigration centers without important industries. Now the industrialization process of the coast will become self-intensifying without a definite regional planning, to the detriment of the rest of Algeria. The processing chain "crude oils-olefines-chemical intermediate products-end products" brings about close technical and locational relations between the mineral-oil-processing, petrochemical, and chemical industries, reinforcing the locational quality of the emplacement once chosen. The emerging chemical industry, which is at the same time an energy and crude-oil consumer, belongs to the key industries that have especially strong effects of agglomerating other industrial enterprises.

So it makes sense that the political integration of the oil production into the national economy is followed by the development of oil-processing industries.[2] Algeria belongs without a doubt to those oil-producing countries in which the development of industries is possible and reasonable. The prerequisites exist to absorb the incomes from oil exports: sufficiently large territory and population and the resulting work force potential, additional resources helpful to industrialization, and existing material infrastructures.

The concentration of development on a few growth poles at the coast was almost inevitable with regard to the most economic use of the scarce capital available. However, it also caused a host of problems. In the new industrial cities like Arzew and Skikda, the capacities of the previously existing infrastructures (roads, dwellings, communication and transport facilities, water supply and sewage system, health, recreational and cultural facilities) are more than exhausted. This leads to social tensions and economic friction losses.[3] These tensions will increase even more if the still-growing social and economic gap between the cities and rural areas is not narrowed. Whereas in the cities a new well-paid and self-conscious class of industrial workers is developing, rural areas are only partially affected

by transformations coming along with the industrial revolution desired by the government. The abrogation of this recognized dualism
and the initiation of mutual learning processes is the goal of the bri-
gades of young volunteers from the cities who advise the traditional
rural population within the actual "agrarian revolution."[4] The large
industrial installations on the coast of the public sector have mainly
pilot functions. An active employment policy seems to be the task
of the private sector. It is the latter that should improve the employment rate of actually only 15 percent of the total population (considering also the rural underemployment).[5]

Ghardaia (private sector), Setif (public sector), and El Asnam
(public sector) are examples of employment creation outside the
coastal regions. In El Asnam, 6,200 persons are to be employed
in the future in the plastics-processing industry. The programs
against the regional disparities (programmes spéciaux, see Chapter
3) also show that the Algerian government is aware of and combating
the problems of this dualism. "For the Algerian government, the
basic industries are just a starting point, a development nucleus,
around which associated industries are to be grouped. They are to
induce new industrial activities by backward and forward linkages
and so to trigger an industrial chain reaction."[6]

The development of the road and railroad networks will facilitate, at least within North Algeria, regional diversification. All of
the projects whose location is not fixed by raw material demands
and transport infrastructures (for example, textile plants in Kabylia,
the tractor plant in Constantine, the plastics-processing industries
in Setif and El Asnam) are outside the favorite regions and intentionally
placed in regions with particularly small developing potential and
strong population pressure.

Effects on Population Patterns

Unfortunately, no comprehensive data are available of the population migrations and shifts caused by the development of new industrial centers. The analysis of population development in the individual
Wilayas indicates the attraction of the industrializing regions (see
Chapter 3). However, because of the size of the statistical units
(Wilayas), which at the same time comprise immigration and emigration regions, it is impossible to give exact figures. The above-average population increase of the new industrial centers is dealt with in
our case studies, but it is not possible to compare them on a macroregional level.

Statistics show that, at least in Algeria, industries related to
hydrocarbons production and processing have the highest employment

growth rate. Employment grew, in the hydrocarbons sector, from
8,578 (1968) to 39,663 (1974) persons—the highest growth rate among
all branches. This, however, was partly occasioned by the integration of former contractors. The number of employees in the chemical
industry also increased above average from 7,121 to 12,562 (76.4 percent; Algerian average of all industrial branches: 62.1 percent).

Absolute data show that most of the new employment opportunities
have been and are still created in the traditional centers of industrial
development such as Algiers. Of the 190,077 (43,728) industrial jobs
created in Algeria from 1969 to 1973 (figures in parentheses refer to
1972-73), 36.3 (44) percent were situated in Algiers, 10.4 (13.7)
percent in the Oases Wilaya, 10.8 (12.3) percent in Oran Wilaya,
which includes Arzew, 15.8 (4.9) percent in Annaba Wilaya (including
El Hadjar steel works and SONATRACH phosphate fertilizer industry),
and 50 (4.3) percent in Constantine Wilaya. The remaining 11 Wilayas
offered only 31.4 (20.8) percent of the available new jobs.[7]

The social aspect of the "innovations" caused by the oil industries (changes in motivation and social behavior) already has been
shown in the regional chapters. Here, too, the oil industry and its
associated industries have pilot functions. Table 11.1 indicates the
demand of workers with above-average qualifications. The advance
in education, which is paralleled by advantages in income and pres-
tige, is intensified by the education and training within the companies.
For instance, in the artificial fertilizer plant in Arzew in two years
268 persons (50 percent workers) were registered in the training and
education facilities within and outside the company.[8] They make up
45 percent of the entire work force of 596 persons. The result was
a 99 percent "Algerianization" of the plant by 1970.

With the introduction of industrial production forms from Western industrialized states a new group of industrial workers is emerging. They are easily distinguished from those active in traditional
trades and agriculture. Their contacts with a foreign culture, technology, and work procedures bring about specific behavioral patterns
manifested in acceptance of modern technology and work procedures;
complete dissociation from the traditional spheres of work; understanding of a modern school education as necessary for children of
both sexes in view of their professional advancement; cultural estrangement from the traditional background, which is, however, less
intensive than among workers emigrated to Europe and students
studying abroad; persistence, within the family life, of traditional
behavioral patterns, especially among first-generation industrial
workers.[9] At least for the second-generation industrial workers with
a modern and mother-tongue education it may be possible to bridge
the split between Arabic-Islamic culture and Western technology by
accepting new technologies without their colonial or foreign background.

TABLE 11.1

Qualifications of the Algerian Industrial Workers in Hydrocarbons,
Chemical Industry, and Other Branches, Including Construction
Enterprises, 1970 and April 1974

Field	Oil	Chemistry	All Branches
Size of work force			
1970	16,240	7,411	310,106
1974	39,633	12,562	448,197
Percent unskilled laborers			
1970	15.6	28.8	34.6
1974	28.9*	25.2	30.6
Percent trained			
1970	22.9	29.4	18.7
1974	17.9	25.5	19.8
Percent specialists			
1970	49.3	32.2	39.8
1974	31.3	40.8	40.5
Percent foremen, technicians			
1970	6.5	7.1	4.5
1974	12.8	5.0	5.5
Percent executives			
1970	5.7	3.3	2.4
1974	8.9	3.4	3.5

*A Result of contracting personnel integration.

Sources: Annuaire statistique de l'Algérie 1972 (Algiers, 1973),
p. 37; and "Les resultats (partels) de l'enquête emploi et salaires de
1974" (Algiers: Sécrétariat d'Etat au Plan, 1975) (mimeographed).

It seems that this increasing economic and social dualism is, for a
limited time at least, inevitable. According to Algerian planning authorities, the solution of the unemployment problem—which has high
priorities—lies in overall industrialization. They believe that this
requires a capital goods industry and is not merely achieved by developing import-substituting light industries.

Algeria hopes in the long run to create more jobs with this industrialization policy than could be provided through other growth strategies (for instance, development of labor-intensive industries). But
the country must realize that for a while the high investments in capital-intensive industries will only contribute slightly to solving the se-

rious unemployment problem. The country must also be aware that the disparities between towns and rural areas as well as those between the traditional and modern working world will increase.

Even in the capital-goods industries, Algeria predominantly decided on the most modern and thus capital-intensive production techniques, when alternative production processes were possible. The far-reaching automation in the new production units therefore results not only from the lack of qualified specialists but, moreover, from the country's desire to quickly catch up technologically with the industrialized world. The "industrial revolution" will presumably affect 10 million Algerians by 1980;[10] however, that will be less than half of the then-expected population. Therefore, it is especially important to integrate the rural and traditionally employed population into the development and modernization processes.

EFFECTS ON REGIONAL ECONOMY
AND AGRICULTURE

The development of a heavy industry based on oil and natural gas in Algeria cannot be looked upon as a process isolated from other fields of the economy.

An Arab [that is, national] world policy which credits hydrocarbons with the role in the development process which they deserve, should not be restricted to a maximization of incomes, the control on the oil investments, or even the construction of several refineries or petro-chemical plants. A policy of this kind must rather bring about a transformation of the total economic environment and a systematic increase in connections between hydrocarbons and other sectors of the national economy as well as promote growth stimuli between these. Beyond monetary transactions, real flows have to be stimulated which enable a maximal exchange of goods and services between oil/hydrocarbons centers and all the other forward and backward linked sectors of the economy. Only by these efforts can the dualism and the isolation, which are still characteristic of the oil industry in Arab countries, be abolished and an accentuated and balanced growth of the national economies be secured. [11]

Nevertheless, the fact that only the basic industry was developed during the first years of industrialization suggests a world market intensive production. But, aside from the natural gas liquefaction plants and the refineries (the latter are already covering the domestic need), the first SONATRACH petrochemical projects were the nitrogenous fertilizer fabrication plant in Arzew (under construction from 1967 to 1969) and the artificial fertilizer fabrication plant in Annaba utilizing the phosphates of the Djebel Onk (planned since 1966, completed in 1971). Whereas during the planning period it was still uncertain whether the two units would produce for the domestic or foreign markets,[12] the national agriculture absorbed the entire output after production started. In fact, the Algerian development concept attaches special importance to a close linkage between agriculture and the industries based on oil. The growing industrial work force with an increasing nutritional demand has to be supplied. At the same time the industries themselves exert impacts on agriculture. The oil and petrochemical industries will do the following in this regard: supply the necessary energy in the form of fuel and electric power; raise its productivity by providing artificial fertilizers and means of pest control; and fabricate plastic items (packing material, pipes, foils, tools) that facilitate agricultural work and the shipment of products.

The Algerian development strategists see no value in exporting crude oil and then purchasing the end products from foreign countries.[13] A detailed study ordered by SONATRACH[14] elaborated again the various levels of connections between the petrochemical industry and agriculture. These links become effective not only at the level of the markets, where the demand for agricultural products rises as a result of increasing purchasing power of the industrial workers. For agriculture, industrial inputs may also become cheaper because of import substitution. In fact, the connections are still closer:

The agricultural production processes are being basically modernized. The industrial products stimulate agricultural productivity directly (artificial fertilizers, means of pest control) or indirectly (foils, pipes, tools). At the same time, new agricultural techniques are introduced and made possible by the products of the petrochemical industry. This is accompanied by a change in the structure of agriculture itself (agrarian reforms, mergers into cooperatives) and of the distribution flows of the forward- and backward-linked products.

The exchanges are not only restricted to the petrochemical industry and agriculture sectors. Corresponding to the increase and mechanization of agricultural production is the increase in the demand of products from the metal-processing industry, for instance. Together with the growth of the latter's production for agriculture, it can at the same time increase its capacity for the supply of other economic fields. This stimulates an overall and integrated development process ranging from the fabrication of the simplest tools to the delivery of complex industrial equipment. This development process

encompasses all of the economy and is not exclusively based on the
links between agriculture and the (petrochemical) industry; however,
this connection plays a key role in this process.

In a colonial-type economy the production surpluses as well
as those of the mining industry had to be exported more or less un-
processed. Thus the producing country was deprived of the eventual
added value resulting from further processing. Only the integration
of all economic sectors into the framework of national planning enables
the use of these production surpluses for the most various economic
sectors, starting a continuously self-intensifying process of producing (added) value. Developing structures have to be assessed in the
overall context, and it follows that in the current stage of industrialization each individual project depends not on short-term profit but rather on the middle- and long-range benefits.

It is difficult to make our rather general statements operational
on a microregional level. This is because of the large number of
factors responsible for regional development (of which the oil industry is only one, although an important, factor), the lack of statistical
data or detailed surveys, and the short time elapsed since operations
began in the industrial plants. However, it can be concluded from the
case studies that the integration of all economic sectors has not yet
been achieved on the regional level. So far, the dualism between the
modern and traditional sectors as well as the problems of agriculture
have been quite aggravated. This is especially visible in the emigration movement of the qualified workers from regions of a comparatively
high level of development, and also by the shortcomings of supply,
which sometimes occur in the industrial centers.

The difficulties brought about by the poorly developed infrastructures in the emerging cities were previously mentioned. However,
trade and other private sectors certainly profit from this "boom."
But a better adjustment to new needs is generally impeded by the lack
of entrepreneurs in the productive sector and by the limiting nature
of the nominally socialist concept of economic development. As seen
in the rural areas, the individual savings are first of all invested in
the construction of a house. In contrast, in urban centers, personal
savings are used for private consumption rather than house construction or productive investments. This results from the lack of space
and construction enterprises as well as from the higher mobility of
the industrial workers. Because a direct link between savings and
productive investments is actually lacking, the government tries to
channel private savings into the nationalized banking apparatus so that
they in turn may be used for the country's productive investments.

The difference between theory and practice of integrated economic planning by the government is evident. Nevertheless, today

the Algerian model of development seems to have good chances for realization and may prove to be an example for the developing nations of the Third World.

NOTES

1. Ch. Verlaque, "L'industrialisation des ports méditéranéens de l'Afrique du Nord," Maghreb et Sahara—Etudes géographiques offertes à Jean Despois (Paris, 1973), pp. 391-413.

2. "Après les hydrocarbures, la pétrochimie part en flèche," Europe Outremer, no. 533/534 (1974), pp. 23-26.

3. See R. Benattig, "La SONATRACH à la veille du 2ème plan quadriennal," Algérie Actualités, no. 438 (March 10-16, 1974).

4. See, for instance, B. P. Balta, "L'est Algérien a l'heure du 'boom' industriel—Les retombées sociologiques," Le Monde, April 20, 1974, p. 10; and Keith Sutton, "The Agrarian Reform in Algeria—The Conversion of Projects into Action," Afrika Spectrum (Hamburg), no. 1 (1974), pp. 50-68.

5. According to J. Franchet, "La formulation de l'éspace algérienne," Annuaire de l'Afrique du Nord 1972 (Paris, 1973), pp. 39-53.

6. According to A. Arnold, "Die Industrialisierung in Tunesien und Algerien," Geographische Rundschau 23, no. 8 (1971), p. 309.

7. Annuaire statistique de l'Algérie, 1974 (Algiers, 1975), pp. 40-41.

8. Following M. Mokrane, "Communication sur la formation des cadres à la SONATRACH," Revue algérienne des Sciences juridiques économiques et politiques, no. 2 (1971), pp. 451-57.

9. See K. Schliephake, "Industriearbeiter und Kulturkontakt in Algerien und Tunesien," Vierteljahresbericht FES—Probleme der Entwicklungsländer, no. 63 (March 1976), pp. 49-62.

10. According to Franchet, "La formulation de l'espace algérien," p. 43.

11. According to Nicolas Sarkis, "L'intégration des hydrocarbures aux économies arabes," Revue algérienne des Sciences juridiques économiques et politiques, no. 2 (1971), p. 501.

12. For this question see, for instance, "Le marché algérien—Le pétrole, le gaz et leur prolongements industriels," Marchés tropicaux, no. 1231 (June 14, 1969), pp. 1665-73.

13. According to A. Ghozali and G. Destanne de Bernis, "Les hydrocarbures et l'industrialisation de l'Algérie," Revue algérienne des Sciences juridiques économiques et politiques, no. 1 (1969), pp. 267-71.

14. See the concise summary by L. Dessau, "L'agriculture et l'industrie pétrochimique," Le Pétrole et Gaz arabes, no. 39 (1970), pp. 38-44.

12

EFFECTS ON FOREIGN
RELATIONS AND ·
COMMERCE

In the previous chapters, questions concerning natural gas production, transportation, and utilization were not thoroughly dealt with, except for the Hassi R'Mel case study (Chapter 6). It is rather difficult to show the spatial aspects of the relatively young natural gas industry, which is extremely capital intensive and labor extensive.[1]

GAS RESERVES AND PRODUCTION

Algeria has 5.6 percent of the world's natural gas and 60.8 percent of Africa's reserves (see Table 12.1). Officially, actual reserves of 3,570 billion cubic meters of gas are now assumed.

Those quantities of this hardly tapped energy source, which cannot be utilized domestically, should be exported. They can thus help to obtain the necessary means for an accelerated economic growth of the country. The use of natural gas creates technical transport and processing problems, because the gas volume must be reduced by extreme cooling (liquefaction) in order to be transported by ship. In 1962/63 preparation for gas liquefaction began, enabling export in 1964 from the CAMEL plant in Arzew. In the meantime, more plants were constructed in Skikda and Arzew (see Chapter 10). An additional plant is projected east of Algiers between Cap Djinet and the Oued (river) Isser. Its capacity should be 13 billion cubic meters per year (with plans for extension to 21 billion cubic meters), but the construction has not yet been contracted.[2]

TABLE 12.1

Algerian Natural Gas Reserves
(in $10^9 m^3$)

Proven Reserves	Amount
Hassi R' Mel	2,000
El Biod region	375
Illizi-Bassin	50
Ahnet, Timimoun, and Reggane regions	375
Total proven reserves	2,800
Almost surely proven reserves	800
Probable reserves	2,900
Possible reserves	600
All reserves together	7,100

Source: N. Ait-Laoussine, "Les exportations algeriennes de gaz naturel," Petrole et gaz arabes (Beirut), no. 127 (July 1, 1974), p. 31.

GAS EXPORTS

SONATRACH has the domestic monopoly of natural gas production, transport, and processing. It ships the liquefied gas together with foreign enterprises to European and U.S. harbors. Its tanker fleet will have, in 1979, a capacity of 674,000 cubic meters of LNG. In 1973 a total of 5,621 million cubic meters of gas were extracted, of which 4.035 million cubic meters were exported.

The capacity of the gas liquefaction plants cannot keep up with the ambitious gas exporting projects that are planned as shown in Table 12.2 and for which some definite and some precontracts were signed. According to the contracts listed, a yearly average of 5.4 billion cubic meters of gas are to be exported; additional contracts are signed for another 27 billion cubic meters and precontracts yet to be approved concern 46.2 billion cubic meters. For instance, SONATRACH and the U.S. Tenneco enterprise contracted in 1976 (subject to government approval), beginning with 1981, the delivery of 10 billion cubic meters of LNG annually for the U.S. and Canadian markets. Once all contracts and precontracts are fulfilled, from 1981 on 70 billion cubic meters of gas will be exported, requiring gas production of 85 billion at the fields. With domestic consumption included (see Chapter 13), the total production will then attain 100 billion cubic meters.[3]

TABLE 12.2

Algerian Export Contracts for Liquefied Natural Gas
as of December 1975

Company	Billion Cubic Meters per Year	Duration in Years	Beginning of Exportation
British Gas	1.0	15	1964
Gaz de France	0.5	15	1965
Gaz de France	3.5	15	1972
Gaz natural (Spain)	1.5	15	1974
European Consortium[a,b]	15.5	20	1980
Snam/ENT (Italy)[b]	11.75	25	1978
Enagas (Spain)	4.5	20	1976/80
Ruhrgas (Germany)[b]	12.0	20	1979
Europe total	50.25		
Distrigas	0.4	20	1971
Distrigas	1.2	20	1975
El Paso Natural Gas I	10.0	25	1976
El Paso Natural Gas II[b]	10.0	20	1980
Eascogas	6.0	23	1977/79
Panhandle	4.5	20	1980
U.S. total	32.1		
World total	82.35		

[a]The European Consortium consists of Gaz de France, 3 billion cubic meters per year; Distrigaz (Belgium), 3 billion; Saar-Ferngas (Germany), 3 billion; Bayern-Ferngas (Germany), 2 billion; Austria-Gas, 2 billion; Gasversorgung Süddeutschland (Germany), 1.5 billion; Swissgaz, 1 billion. As the negotiations with the Consortium have not yet brought any results, separate contracts have been concluded with SONATRACH by (1) Belgian Distrigaz in November 1975, effective since June 1976. From 1979 on, SONATRACH will deliver annually for 20 years 3.5 billion cubic meters of LNG from its Arzew plant. (2) French Gaz de France on April 2, 1976, covering also 3.5 billion cubic meters to be delivered during 20 years from 1980 on. Algerian sources indicated that prices will be between $1.30 and $1.40. See "Accord SONATRACH-Gaz de France," Maghreb-Machrek, no. 72 (1976), pp. 15-16; and Industries et Travaux d'Outremer, no. 272 (July 1976), p. 540.

[b]Precontracts partly nonapproved.

Sources: "L'Algerie, premier exportateur mondial de gaz naturel avant 1980," Europe Outremer, no. 533/539 (1974), pp. 19-22; Petroleum Economist, February 1975, p. 51; and December 1975, p. 455.

Gas transport to Italy according to the contract with the Italian national ENI enterprise was originally to be conducted through an underwater gas pipeline reaching the Italian mainland via Tunisia, Sicily, and the Strait of Messina. At a length of 2,600 kilometers and maximal depths of 500 meters it was to transport 11 billion cubic meters annually (70 percent of the Italian consumption in 1972) and to become operational in 1978. Technical investigations in 1971 had shown the feasibility of this project and of a similar pipeline going to Spain via Morocco and the Strait of Gibraltar. Total costs of both projects were estimated at $10 billion. However, these plans did not materialize because of difficulties with Algeria's neighbors, which barred the transit routes or demanded high transit duties. A new contract between SONATRACH and ENI supersedes the previous one and schedules annual exports of 8.5 billion cubic meters to Italy via tankers from Skikda. With government approval obtained, shipment may begin in 1981 and will last 20 years. [4]

Prices listed in contracts prior to the energy crisis were about $.40 to $.80 per million British Thermal Units (BTUs); prices listed in 1975 and 1976 contracts approximated $1.30 to $1.40 (plus transport) and are indexed in relation to fuel oil prices in New York and on the U.S. Gulf Coast as well as in relation to the performance of the dollar against a group of European currencies.

The secondary, political results of such contracts are a close economic interrelation between producer and consumer countries. Only certain industrialized countries with high energy demands and developed networks of gas pipelines can appear as buyers. With the construction of capital-intensive transport and liquefaction utilities, which demand stable and continuous relations between seller and buyer, Algeria will become their long-term partner. Thus, natural-gas exporting will narrow the margin in foreign trade and politics of the Algerian government. On the other hand, once all contracts are carried out according to the Algerian concept, the state will expect yearly revenues of approximately $230 million from gas export by the end of 1979 (see also Chapter 3). In the long range the value of natural gas exports should reach the same level as the oil exports.

It also should be noted that a certain pride has led the Algerians to develop an industry with such a high technological level as is the case of natural gas liquefaction and transport. The gas liquefaction plant pioneered in 1964 in Arzew has since served as a school, as Algeria is still dependent on imported technology and equipment. Thus, the country already has a highly specialized labor potential for this sector.

Just like all of the technologically advanced heavy industries, natural gas handling led to the education of a technical elite, who in turn transmit and propagate their knowledge and motivation. "The factories and facilities cost us a lot of money, but the most significant profit is the knowledge which the workers in these facilities have acquired and which turns the technological progress into their own."[5]

NOTES

1. The following data are from numerous publications in the special literature, for instance "Où en sont les contrats algéro-americains sur le gaz," Pétrole Informations (Paris), December 7-13, 1973, pp. 23-24; "L'Algérie commande en France 5 méthaniers de 125,000 m^3," Pétrole Informations, January 18-24, 1974, p. 23; "Algerian-Moroccan Gas Pipeline," BBC Summary of World Broadcasts, ME/4554/A/2, March 19, 1974; "L'ENI commence les travaux du gazoduc Algérie-Italie," Pétrole et gaz arabes (Beirut), no. 122 (April 16, 1974), pp. 14-15; "Le GNL facteur de développement de l'Algérie (interview with A. Ghozali)," Pétrole Informations, June 14-20, 1974; "RFA/Algérie: un accord solide," Pétrole Informations, June 21-27, 1974; "Regards sur l'actualité—Algérie," Pétrole et gaz arabes, no. 127 (July 1, 1974). A general survey is given in "L'approvisionnement en gaz naturel de l'Europe et de la France," Maghreb (Paris), no. 54 (1972), pp. 40-45; "Le gaz naturel, principal atout de l'Algérie?" Maghreb-Machrek, no. 62 (1973), pp. 21-34; and Nicole Grimaud, "Essor de la commercialisation du gaz naturel algérien," Maghreb-Machrek, no. 71 (1976).

2. The Industrial Revolution (Algiers: Ministry of Information and Culture, 1976), p. 52.

3. Moudjahid, October 5, 1976.

4. According to an ENI Press Release dated December 29, 1976. For the older project, see also Angelo Rocca, "Aspects quantitatifs de la commercialisation du gaz naturel arabe en Europe jusqu'en 1980," Revue algérienne des Sciences juridiques économiques et politiques, no. 2 (1971), pp. 507, 513.

5. A. Ghozali, "Le GNL facteur de développement de l'Algérie," Pétrole Informations, June 14-20, 1974, p. 26.

13

It is the target of the Algerian "natural gas strategy" to use the gas as an energy source for industrial and household purposes as well as raw material for industrial processing.[1] According to these directives, the following has been achieved:

The installation of a national network of gas pipes that today supplies the most important cities and industrial plants. Within a few years, 3,000 kilometers of pipes were laid to increase the domestic gas use.

The use of natural gas in the Arzew plant for the fabrication of ammonia, nitrogenic fertilizer, and synthetic resins. A petrochemical complex is under construction in Skikda using mainly the natural gas product ethane.

The extraction of LPG associated with the oil of Hassi Messaoud, its transportation, and its processing for export in Arzew.

INTERNAL CONSUMPTION

These measures brought about an increase in the domestic consumption of 300 million cubic meters in 1962 to more than 2 billion cubic meters in 1973. As a result of the use of natural gas (as a raw material) for the fabrication of ammonia and other petrochemical products, the consumption is expected to quadruple until 1977 and to reach 12.5 billion cubic meters in 1981.[2]

The entire population benefits from the inexpensive energy source provided by natural gas. Already in 1968 the price of domestic gas was reduced by 50 percent. Whereas 18 cities (consuming 430 million cubic meters) were supplied in 1968, the Four-Year Plan projected the supply of another 53 cities by 1973 that should have brought 120,000 households into the system. During the plan period, investments of DA 200 million were necessary. For instance, a 300-kilometer feeder pipeline, completed in 1974, branches off the Hassi R'Mel-Skikda pipe in Oued Athmania. It supplies Setif, Bejaia, and other cities of the then Wilaya Setif.[3] The construction cost of the pipeline totaled DA 80 million. In the town of El Eulma, for instance, a 60-kilometer network of city gas pipelines was assembled, at a cost of DA 5 million, connecting 3,500 households. During the Second Four-Year Plan 1974-77, another 44 localities were linked to the gas network by the help of 700 kilometers of main pipes.[4] Besides the industrial enterprises (such as the nuts and bolts factory in Ain Kebira and the brickyards of Bordj Bou Arreridj) the households are also supplied with their energy needs.

This will also contribute to the preservation of the natural vegetation. For instance, the destruction of forests by firewood gatherers can be halted if appliances and fuels for kitchen and heating purposes are also available for low-income households.[5] In fact, a certain success has been achieved in rural areas. Sari reports from the mountainous Ouarsenis region that the use of bottled natural gas as a domestic energy source is becoming more and more common since the price for bottled gas was reduced from DA 15 to DA 10 per bottle.[6] However, in order to preserve the forest reserves, a further price reduction of bottled gas, furnaces, and gas stoves has to be ensured.

PRODUCTION OF ELECTRICITY

As the export of LNG is somewhat costly and risky (see Chapter 12), gas is used more and more for the production of electric energy in Algeria and Tunisia. In Algeria, besides the gas turbines installed in or near the oil/gas fields (see Chapter 6), additional natural-gas-powered generators were commissioned to be built in the north, in 1975. Additional production capacity to be achieved during the Second Four-Year Plan 1974-77 amounts to 1,510 megawatts, distributed as follows:[7]

	Megawatts
Hassi Messaoud	112
Hassi R'Mel	90
Boufarik (near Algiers)	100
Bab Ezzouar (near Algiers)	120
Skikda	270
Jijel	600

	Megawatts
M'Sila	200
Diesel powered (in the south)	42

Nearly all of these new plants will be powered by natural gas.

In Tunisia, gas with estimated reserves of 60 billion cubic meters will also be sufficient to cover the country's industrial and individual energy consumption, at least until 1995. It will thus enter progressively into electricity production. [8] Of the total national electricity producing capacity in 1975 (345 megawatts), 48 percent were powered by natural gas (see, for instance, the Gabes section in Chapter 9), 9 percent by water power, and the rest by oil. In STEG's (the national electricity and gas company) development program to provide an additional 415 megawatts, gas will account for the generation of at least 90 megawatts. [9]

Subsequently it will be the task of the Algerian and Tunisian natural gas supplies to satisfy the countries' energy demands that are rapidly increasing as in other developing countries. As the overseas handling and transportation of oil is comparatively easier than that of natural gas, it would certainly make sense in a future energy strategy to substitute, on a long-term basis, domestic oil consumption by natural gas. Thus oil deposits may be spared and/or more oil will be at the disposal for export bringing in foreign currency.

NOTES

1. See, for instance, N. Ait-Laoussine, "Les exportations algériennes de gaz naturel," Pétrole et gaz arabes, no. 127 (July 1, 1974), pp. 32-34; "Electricité et gaz pour tous," Algérie et Développement, no. 5 (January-February 1971), pp. 27-29; as well as quoted literature.

2. After Moudjahid, October 5, 1976.

3. According to "Natural gas distribution in Algeria," BBC Summary of World Broadcasts, ME/4554/A/2, March 19, 1974; and "Par-ci, par-là," Moudjahid, May 5-6, 1974.

4. See "Stand SONELGAZ—Des chiffres qui en disent long," Moudjahid, September 10, 1975.

5. A. Ghozali and G. Destanne de Bernis, "Les hydrocarbures et l'industrialisation de l'Algérie," Revue algérienne des Sciences juridiques économiques et politiques, no. 1 (1969), p. 271.

6. Dj. Sari, "La désorganisation de l'agriculture traditionelle dans l'Ouarsenis," Etudes rurales (Paris), no. 47 (1972), p. 60.

7. See Industries et Travaux d'Outremer, April 1976, pp. 296; and "Stand SONELGAZ." More recent data indicate a capacity of 600 megawatts for M'Sila, to be used within the planned aluminum smelter. The smelter, to be delivered in 1982 by the Soviet Union, will produce 140,000 tons of aluminum annually, and will be powered exclusively by natural gas (approximately 800 million cubic meters annually). Following "L'U.R.S.S. va construire une usine d'aluminium à M'Sila," Marchés tropicaux, no. 1591 (May 7, 1976), p. 1206.

8. Marchés tropicaux, no. 1539 (May 9, 1975), p. 1362.

9. "Le programme de développement de la STEG," Marchés tropicaux, December 26, 1975, p. 3676.

The energy crisis of the fall of 1973 was only superficially caused by the flare-up of the Palestine conflict. The main reason was the desire of the Arab oil-producing countries to gain full control of this very important raw material and at the same time to secure the highest possible price. This desire did not develop spontaneously, but was the result of intensive investigation and evaluation by the oil-producing countries of the economic and social processes associated with the oil industry. These processes develop within triangular interrelationships of natural resources, economic mechanisms, and political directives connected by many interdependencies. To analyze these interrelationships is certainly one of the main tasks of science in developing countries. The result of these examinations can explain political actions and thus provide guidelines for future concepts.

The effects of the oil industry are twofold: On the one hand, it provides the producer country with high monetary income. These effects are called indirect or secondary (as the state is a catalyst and no direct and automatic functional relationship exists between the origin and the actual use of the capital). On the other hand, there are direct, primary effects on the regional economy such as new jobs, improvement in the infrastructure, and impacts on existing or developing forward- and backward-linked commercial and industrial activities.

In contrast to the economist's usual approach, directed more toward the indirect effects, our study analyzes the direct, regional effects and impacts by the means of microregional case studies.

An overall view of the natural settings and the social and economic structures shows three problems that are basically similar in the two countries (even though differences in standards exist):

The regional disparities. The coastal zones in the north of each country are starting points and locations for a modern economic development and industrialization. In contrast, the southern mountain,
 steppe, and desert regions are in a marginal situation due to their
 unfavorable natural setting and their transport isolation.
The deficiency of agriculture, mainly in Algeria. In spite of all modernization attempts it is still divided into two clearly distinguish-
 able sectors. Whereas the modern sector is rather export oriented,
 the traditional sector hardly nourishes its own work force.
The lack of jobs in the trade and industries sector. These have to be
 gradually created within the industrialization programs now becoming effective.

With these problems in mind, a survey of the history, the current importance of the oil industry, and the energy policies and strategies clearly shows (1) the difference in size of the oil industry in the
two countries (Algeria's annual production is around 50 million tons,
Tunisia's is 5 million tons); (2) the entirely different approach to an
eventual energy and raw material strategy. Algeria, basing its concept on its regional experiences and evaluations (shown in our regional chapters), pursues a strategy to use oil as the development basis of the domestic industry as well as a power factor in foreign policy. Tunisia's oil concept was formulated only a very short time ago.

The regional survey of the spatial effects of oil production,
transportation, and processing yields the following tendencies and potentials:

Oil production in the Algerian Wilaya Oases monetarily is the
most important activity. In Tunisian producing regions, its effects
are visible, but not very important. Further and long-lasting impacts cannot be expected because of the fugitiveness and transitori-
ness of all mining and the still-lacking integration of the oil sector
into other fields of the regional economy, and also because of the
natural setting and the marginal location of the producing regions.
If the respective oil production decreases significantly, the oil settlements probably will be deserted soon.

The transport facilities (mainly the oil and natural gas pipelines)
are too capital intensive and labor extensive, and also technologically
too specialized, to cause regional effects. Instead, their end terminals in the shipping harbors are locations favored by their transport
situation and their equipment. Also, the necessary personnel and material infrastructures are generally provided here enabling industries
to develop. Such industries may be based on oil/natural gas as a
cheap energy source and/or raw material for the petrochemical industry (producing fuels, fertilizers, plastics, chemicals, and so on).

Therefore, the genuine, strong regional impacts on the labor market and on possible forward- and backward-linked sectors of industry and trade are generally most effective in the seaport towns— but only when they are integrated into the framework of national planning that recognizes their potential. Integrated (Arzew, Skikda, as special cases also Gabes and Setif) and nonintegrated (Bejaia, Skhira) harbors are analyzed in the regional chapters.

Here backward linkages of "secondary" with "primary" effects also occur. Without the necessary revenues from the crude oil export, the producing countries would not be able financially to develop industries and so to utilize the advantages of these locations. Such utilization is indispensable for the survival of these countries. As an exhaustion of the oil reserves can be expected in one or two generations, on the basis of the current producing rhythm, a certain qualified degree of industrialization and economic independence has to be achieved by that time. It is obvious that the raw materials of the mining industry should be the base of this policy. The margin of action with regard to foreign trade is being extended. But the national policy concerning resources becomes fixed because the production has to be controlled so that the new petrochemical industry can be supplied for a long time. The processed products will have higher values and therefore will certainly more than compensate for the generally sinking incomes from the decreasing crude oil export. The times of cheap energy imports for the already industrialized countries are certainly over once and for all. In contrast, new partnerships of an industrial cooperation are developing in North Africa, and they appear to be reasonable and profitable for both sides.

BIBLIOGRAPHY

Alber, O. "Erdöl und Erdgas in der Sahara." Afrika Spectrum (Hamburg), no. 3 (1967), pp. 38-47.

Aliev, M., et al. Geological Structures and Estimation of Oil and Gas in the Sahara in Algerie. Algiers: SONATRACH, 1971.

Akhtarekhavari, Farid. The Oil Price Policy of the OPEC Countries. Probleme der Weltwirtschaft, Diskussionsbeiträge, no. 2. Munich, 1975.

Amin, Samir. L'économie du Maghreb. Grands documents, no. 25. Paris: Editions du Minuit, 1966.

Amrane, D. Dj. "L'emploi à Bajaia." University of Algiers, Thesis of the Third Cycle, 1970. (Mimeographed.)

Arnold, Adolf. "Die Industrialisierung in Tunesien und Algerien." Geographische Rundschau 23, no. 8 (1971): 306-16.

Attia, H., and M. Rouissi. "Structures agraires et développement dans le Sud tunisien." Annales algériennes de Géographie, Special number (Colloque de Ouargla). Algiers, 1972, pp. 15-34.

Basso, J., and J. Touscoz. "Les stratégies maghrébines pour l'exploitation et l'exportation des matières premières—Quelques problèmes de droit international public." Annuaire de l'Afrique du Nord 1974. Paris, 1975, pp. 73-205.

Only the more important items, especially those of a general scientific interest, are listed. Newspaper articles and other titles giving practical information are found in the notes to the corresponding chapters. A comprehensive bibliography has been edited in consultation with the author under the title Oil in Africa, Documentation Service Africa, Series A, vol. 12 (Hamburg: Institut für Afrika-Kunde, 1975). Two other regional bibliographies have been edited by the same institution: Algeria—Economy and Development Planning, vol. 5 (1972); and Tunisia—Economy and Development Planning, vol. 10 (1974). For current bibliographies, especially of the francophone literature, see also Annuaire de l'Afrique du Nord (Paris: CNRS, since 1962).

Benbella, A. "La politique pétrolière algérienne." Revue algérienne des Sciences juridiques économiques et politiques, no. 4 (1964), pp. 83-103.

Ben Slama, Moncef. "Régionalisation et planification du développement." Revue tunisienne des Sciences sociales, no. 2 (1970), pp. 91-118.

Bolz, Reinhardt. Tunisia—Economic and Social Structures and Development (in German). Arbeiten aus dem Institut für Afrika-Forschung, vol. 6. Hamburg, 1977.

Bolz, R., and K. Eitner. Institutions of Development Planning in Tunisia. Documentation Service Africa, Series B, no. 3. Hamburg: Institut für Afrika-Kunde, 1974.

Boutros Ghali, B. "The League of Arab States." In The Organization of African Unity. Edited by El Ayouti. New York: Praeger, 1975, pp. 47-61.

Castevert, C., and M. Cote. "Mise au point sur Hassi Messaoud." Annales algériennes de Géographie, no. 9 (1970), pp. 106-17.

Champault, D. Une oasis du Sahara nord-occidental—Tabelbela. Paris: CNRS, 1969.

Cohen-Hadria, E. "Perspectives économiques tunisiennes." Maghreb-Machrek, no. 71 (1976), pp. 9-10.

Corna Pellegrini, G. Per una geografia delle città pioniere—Hassi Messaoud. Pub. no. 4. Milan: Catholic University of Santo Cuore, 1971.

Corneille, A. Les Chaanbas. Paris: CNRS, 1968.

Dessau, L. "L'agriculture et l'industrie petrochimique." Le Petrole et Gaz arabes (Beirut), no. 39 (1970), pp. 38-44.

Destanne de Bernis, G. "Les industries industrialisantes et les Options algériennes." Revue Tiers-Monde, no. 47 (1971), pp. 545-63.

_____. "Les problèmes pétroliers algériens." Revue algérienne des Sciences juridiques économiques et politiques, no. 4 (1973), pp. 717-58.

_____. "Revenues pétroliers et choix d'un éspace de développement."
Mondes en Développement, no. 18 (1974), pp. 73-99.

Duwaji, Ghazi. Economic Development in Tunisia: The Impact and
Course of Government Planning. New York: Praeger, 1967.

Etienne, Bruno. "La place des matières premières dans la politique
extérieure de l'Algérie." Annuaire de l'Afrique du Nord 1974.
Paris: CNRS, 1975, pp. 71-92.

Fakhfakh, M. "L'influence d'une grande ville sur l'habitat de sa région: Sfax." In Maghreb et Sahara—Etudes géographiques offertes a Jean Despois. Paris, 1973, pp. 137-46.

Farley, Rawle. Planning for Development in Libya. New York:
Praeger, 1971.

Franchet, J. "La formulation de l'espace algérien." Annuaire de
l'Afrique du Nord 1972. Paris: CNRS, 1973, pp. 39-53.

Ghozali, A. "L'association coopérative algéro-francaise: ses buts,
son bilan, ses perspectives." Revue algérienne des Sciences
juridiques économiques et politiques, no. 1 (1969), pp. 165-72.

_____. "Le GNL facteur de développement de l'Algérie." Pétrole
Informations, June 14-20, 1974, pp. 26 ff.

Ghozali, A., and G. Destanne de Bernis. "Les hydrocarbures et
l'industrialisation de l'Algérie." Revue algérienne des Sciences
juridiques économiques et politiques, no. 1 (1969), pp. 253-94.

Gilbert, H. "Le pétrole en Tunisie." Maghreb-Machrek, no. 58
(1973), pp. 37-41.

Grissa, Abdessatar. Agricultural Policies and Employment: Case
Study of Tunisia. Development Centre Studies, Employment
Series, no. 9. Paris: OECD, 1973.

Hullot, J. P. "Le pétrole en Algérie." Revue des Etudes politiques
africaines, no. 47 (April 1969), pp. 18-37.

Josse, Raymond. "Problèmes de mise en valeur du Hoggar et de la
croissance urbaine à Tamanrasset." Cahiers d'Outremer
(Bordeaux), no. 95 (1971), pp. 245-93.

Kuklinski, A. R., ed. Growth Poles and Growth Centers in Regional Planning. UNRISD Regional Planning, vol. 5. Paris-The Hague, 1972.

Le Coz, Jean. "Mutations rurales au Maghreb: Du dualisme agraire à l'aménagement de l'éspace." In Introduction à l'Afrique du Nord contemporaine. Edited by W. K. Ruf et al. Paris: CNRS, 1975, pp. 63-80.

Lerat, S. "Hassi Messaoud." Cahiers d'Outremer, no. 93 (1971), pp. 16-31.

Mabrouki, Azzedine. "L'O.P.E.C. et le Tiers Monde—un même combat." El Djeich (Algiers), no. 143 (1975), pp. 29-34.

Madelin, Henri. Pétrole et politique en Méditerranée occidentale. Cahiers de la Fondation nationale des Sciences politiques, vol. 188. Paris, 1973.

Mahrour, M. "L'occupation du sol et la distribution de la population dans le Sahara algérien." Annales algériennes de Géographie, Special number (Colloque de Ouargla). Algiers, 1972, pp. 6-14.

Marthelot, P. "La révolution de pétrole dans un pays insuffisamment développé: la Libye." Cahiers d'Outremer, no. 18 (1965), pp. 5-31.

McNee, R. B. "Functional Geography of the Firm with an Illustration from the Petroleum Industry." Economic Geography 34 (1958): 321-57.

Mefti, H. "Le transport par pipelines en Algérie." Revue algérienne des Sciences juridiques économiques et politiques, no. 2 (1971), pp. 479-90.

Melamid, Alexander. "Satellization in Iranian Crude Oil Production." Geographical Review (New York) 63 (1973): 27-43.

Mensching, H. Tunesien. Wissenschaftliche Länderkunden, vol. 1. Darmstadt, 1968.

_____. "Le milieu naturel du Maghreb: Questions et limites de la mise en valeur du potentiel agraire." In Introduction à l'Afrique du Nord contemporaine. Edited by W. K. Ruf et al. Paris: CNRS, 1975, pp. 49-61.

Moeller, Ties. Mining and Regional Development in East Africa
    (in German). Afrika Studien, vol. 67. Munich: Weltforum,
    1971.

Mokrane, M. "Communication sur la formation des cadres à la
    SONATRACH." Revue algérienne des Sciences juridiques écon-
    omiques et politiques, no. 2 (1971), pp. 451-57.

Murgue, B. "L'équipement industriel de l'Algérie de 1962 à 1975."
    Industries et Travaux d'Outremer, no. 267 (February 1976), pp.
    84-125.

Negadi, G., D. Tabutin, and J. Vallin. Situation démographique de
    l'Algérie. Dossiers documentaires nos. 19-20. Algiers, 1972.

Nesson, C. "Un domaine autogéré au Sahara. Le domaine Hamlaoui
    Brahim à Ourir." Annales algériennes de Géographie, no. 8
    (1969), pp. 149-62.

Odell, R. R. Oil and World Power: A Geographical Interpretation.
    Harmondsworth, 1970.

Pearson, Scott R. Petroleum and the Nigerian Economy. Stanford,
    Calif., 1970.

Pompei, S. "Problèmes d'urbanisme dans le Sahel." Cahiers de
    Tunisie, nos. 47-48 (1964), pp. 147-63.

Robana, Abderrahmane. The Prospects for an Economic Community
    in North Africa. New York: Praeger, 1973.

Rocca, Angelo. "Aspects quantitatifs de la commercialisation du
    gaz naturel arabe en Europe jusqu'en 1980." Revue algérienne
    des Sciences juridiques économiques et politiques, no. 2 (1971),
    pp. 507-13.

Rouvillois-Brigol, M. "Les transformations de l'Oasis de Ouargla."
    Annales algériennes de Géographie, Special number (Colloque
    de Ouargla). Algiers, 1972, pp. 45-58.

Sari, Dj. "La désorganisation de l'agriculture traditionelle dans
    l'Ouarsenis." Etudes rurales, no. 47 (1972), pp. 39-72.

_____. "L'évolution de l'emploi en Algérie." Maghreb-Machrek,
    no. 69 (1975), pp. 42-50.

_____. "Problèmes démographiques algériens." Maghreb-Machrek, no. 63 (1974), pp. 32-42.

_____. "La récupération et la valorisation des hydrocarbures par l'Etat algérien." Annales algériennes de Géographie, Special number (Colloque de Ouargla). Algiers, 1972, pp. 209-37.

_____. "La restructuration des centres urbains en Algérie." Annuaire de l'Afrique du Nord 1972. Paris, 1973, pp. 55-75.

Sarkis, Nicolas. "L'intégration des hydrocarbures aux économies arabes." Revue algérienne des Sciences juridiques économiques et politiques, no. 2 (1971), pp. 491-506.

Schaetzl, Ludwig. Industrialization in Nigeria: A Spatial Analysis. Afrika-Studien, vol. 81. Munich: Weltforum, 1973.

_____. Petroleum in Nigeria. Ibadan, 1969.

Schliephake, Konrad. Libya—Economic and Social Structures and Development (in German). Arbeiten aus dem Institut für Afrika-Kunde, vol. 3. Hamburg, 1976.

_____. "Changing the Traditional Sector in Algeria's Agriculture." Land Reform (Rome, FAO), no. 1 (1973), pp. 19-28.

_____. "Industriearbeiter und Kulturkontakt in Algerien und Tunesien." Vierteljahresberichte FES—Probleme der Entwicklungsländer, no. 63 (March 1976), pp. 49-62.

_____. "Libyen als Bezugspunkt mediterraner Arbeiterwanderungen-Beispiele aus Tunesien und Malta." Orient (Opladen) 15, no. 3 (1974): 112-15.

Sereni, Jean-Pierre. "La politique algérienne des hydrocarbures." Maghreb, no. 45 (1971), pp. 31-49.

Souidi, M. The Development of the Plastics Industry in Algeria. ID/WG.189/3, November 12, 1974. Vienna: UNIDO, 1974.

Sutton, Keith. "Algeria: Changes in Population Distribution, 1954-66." In Populations of the Middle East and North Africa. Edited by J. I. Clarke and W. B. Fisher. New York, 1972, pp. 373-403.

_____. "Industrialization and Regional Development in a Centrally-Planned Economy—The Case of Algeria." Tijdschrift v. Econ. en Sociale Geografie (Amsterdam), no. 2 (1976), pp. 83–94.

_____. "L'industrie algérienne du gaz et du pétrole—Développements depuis l'indépendance." L'Information géographique (Paris), no. 5 (1970), pp. 212–18.

Tanzer, M. The Political Economy of International Oil and Underdeveloped Countries. Boston, 1970.

Tempelman, G. J. "Tourism in South Tunisia: Developments and Problems in the Djerba-Zarsis Region." Tijdschrift v. Econ. en Sociale Geografie 66 (1975): 21–34.

Tiano, A. "Human Resources Investment and Employment Policy in the Maghreb." In Employment in Africa. Geneva: ILO, 1973, pp. 151–75.

Tuma, Elias H. "Population, Food and Agriculture in the Arab Countries." Middle East Journal, no. 4 (1974), pp. 381–95.

Verlaque, Ch. "L'industrialisation des ports méditerranéens de l'Afrique du Nord." In Maghreb et Sahara—Etudes géographiques offertes a Jean Despois. Paris, 1973, pp. 391–413.

Viratelle, G. L'Algérie algérienne. Coll. Développement et Civilisations. Paris, 1973.

Vlachos, G. "Le régime juridique des hydrocarbures en Algérie." Revue juridique et politique Indépendance et Coopération (Paris) 28, no. 1 (1974): 103–28.

Zartmann, William. Government and Politics in Northern Africa. London, 1964.

Zoeller, A. C. "Algerian Nationalizations: The Legal Issues." Journal of World Trade Law (Twickenham) 6, no. 1 (1972), pp. 33–57.

UNAUTHORED BOOKS AND ARTICLES

Algeria in Numbers 1962–72. Algiers: Ministry of Information and Culture, 1972.

"L'Algérie et les hydrocarbures." Annuaire de l'Afrique du Nord 1965. Paris, 1966, pp. 63-100.

"All Oil Countries Want to Set Up Industries." Petroleum Economist (London), no. 9 (1974), pp. 324-26.

"L'approvisionnement en gaz naturel de l'Europe et de la France." Maghreb, no. 54 (1972), pp. 40-45.

"Arrondissement de Gabes—Réalisations récentes et objectifs du Département pour la durée du Plan quadriennal 1973-76." Revue tunisienne de l'Equipement, no. 8 (1974), pp. 15-26.

"Etude Socio-économique sur l'habitat des agents CFPA en Algérie." Neuilly: Euréquip and CFPA, December 1965. (Mimeographed.)

"Les fondements de la planification algérienne." El Djeich, no. 138 (1974), pp. 75-79.

"Gabes devient une capitale industrielle." Europe France Outremer, no. 513 (October 1972), pp. 36-38.

Gabes 73—D'un centre agricole à un pôle de développement. Tunis: Direction de l'Aménagement du Territoire, 1973.

"Le gaz naturel, principal atout de l'Algérie?" Maghreb-Machrek, no. 62 (1973), pp. 21-34.

Industrial Development Survey, vol. 4 ID 83. New York: United Nations, 1972.

Industrial Promotion in Tunisia. Tunis: Agence de Promotion des Investissements, 1975.

The Industrial Revolution. Algiers: Ministry of Information and Culture, 1976.

"L'industrialisation de l'Afrique—les nouvelles usines et les grands projets en 1975." Industries et Travaux d'Outremer, no. 269 (1976), pp. 233-312.

Memorandum Presented by Algeria to the Conference of OPEC Sovereigns and Heads of State. Algiers, March 1975.

Les Oasis. Algiers: Ministry of Information and Culture, 1970.

Le pétrole, les matières de base et le développement—Mémoire présenté par l'Algérie à l'occasion de la session extraordinaire de l'Assemblée Générale des Nations unies. Algiers, April 1974.

IVe Plan de développement 1973-1976. Tunis, 1973.

"Première carte industrielle de l'Algérie." Industries et Travaux d'Outremer, no. 247 (1974), pp. 546-62.

Report in Investment Promotion Meeting and Technological Consultations for Chemical Industries in Developing Countries. ID/WG.197/5, June 5, 1975. Vienna: UNIDO, 1975.

"Les resultats (partiels) de l'enquête emploi et salaires de 1974." Algiers: Sécrétariat d'Etat au Plan, 1975. (Mimeographed.)

Sfax 72—Problems/perspectives d'aménagement. Tunis: Direction de l'Aménagement du Territorie, n.d.

La situation de l'emploi dans le Gouvernorat de Kasserine. Tunis: Ministère du Plan, 1973.

Tunisia: The Development of the Petroleum Industry. E/CN.14/EP/58, September 24, 1973. New York: UN Economic and Social Council, 1973.

Les villes en Tunisie. Tunis: Direction de l'Aménagement du Territoire, 1971.

ABOUT THE AUTHOR

KONRAD SCHLIEPHAKE is Assistant Professor at the Institute of Geography, Würzburg University, Federal Republic of Germany. He has published three books and numerous scholarly articles in German, focusing heavily on regional development problems in Northern Africa. Before joining the faculty at Würzburg, Dr. Schliephake was a Research Official with the German Institute for African Studies, Hamburg. He has traveled extensively in the Arab countries and has concentrated his field work in Algeria, Tunisia, and Libya.

Dr. Schliephake earned his doctoral degree at Giessen University.

ARAB OIL: Impact on Arab Countries and Global
Implications
        edited by Naiem A. Sherbiny
        and Mark A. Tessler

DEVELOPMENT OF THE IRANIAN OIL INDUSTRY
        Fereidun Fesharaki

OIL IN THE ECONOMIC DEVELOPMENT OF
VENEZUELA
        Jorge Salazar-Carillo